HOLDING THE LINE

HOLDING THE LINE

The Battle of Allegheny Mountain

AND CONFEDERATE DEFENSE OF
THE STAUNTON-PARKERSBURG TURNPIKE, 1861-62

JOE GEIGER, JR.

To a fellow historian
and collector,
Best wishes Mitchell
Joe Geiger, Jr.

Quarrier Press
Charleston, West Virginia

Quarrier Press
Charleston, WV

First Edition

10 9 8 7 6 5 4 3 2 1

Printed in the United States of America

Library of Congress Control Number: 2012939612
ISBN-13: 978-1-891852-83-1
ISBN-10: 1-891852-83-3
Book & cover design: Mark S. Phillips

Front cover: Wartime sketch of Romney, winter of 1861-62, by Alfred R. Waud. Library of Congress, Prints & Photographs Division, Civil War Photographs [LC-DIG-ppmsca-22434.]

Back cover: Map of area of operations, parts of Pocahontas and Randolph counties. Official Records.

Distributed by:

WEST VIRGINIA BOOK COMPANY

West Virginia Book Company
1125 Central Avenue
Charleston, WV 25302
www.wvbookco.com

TABLE OF CONTENTS

Acknowledgements vi
Introduction ix
Chapter One: War Erupts in Western Virginia 1
Chapter Two: A Plan to Pay Respects 19
Chapter Three: The Battle of Greenbrier River 30
Chapter Four: A Lively Month 54
Chapter Five: Stonewall Gets His Way 80
Chapter Six: The Battle of Allegheny Mountain 107
Chapter Seven: Winter Work 140
Chapter Eight: Closing the Show 171
Notes 189
Bibliography 211
Appendix A: Union Casualties, Battle of Greenbrier River 220
Appendix B: Confederate Casualties, Battle of Greenbrier River 223
Appendix C: Union Casualties, Battle of Allegheny Mountain 226
Appendix D: Confederate Casualties, Battle of Allegheny Mountain 232
Appendix E: Poetry 239
Index 256
About the Author 263

ACKNOWLEDGEMENTS

My interest in the 1861 campaign along the Staunton-Parkersburg Turnpike in Pocahontas and Randolph counties was stirred by research done by my late grandfather, the Reverend John Quinn Geiger, who was raised on a farm near Stony Bottom in Pocahontas County. Following his retirement from the ministry, Grandpa wrote a book on his life's experiences entitled *Man From Michael's Mountain*. After the book's publication, he conducted further research on the roots of our family, which revealed that my great-great-great grandfather had three sons who fought in the Civil War, all for the Confederacy. I was intrigued by an entry in Grandpa's research papers which noted that one of the boys, Harman Geiger, died in early January 1862 from wounds received in battle. At first I believed he had been killed at the Battle of Allegheny Mountain, but it appears that he never officially enlisted in the Confederate army and was probably mortally wounded while serving with the militia during the Union raid on Huntersville on January 4, 1862. Once I began to gather materials about the battle, I immersed myself in the campaign along the Staunton-Parkersburg Turnpike following the failed Confederate attack on Cheat Summit Fort in mid-September 1861.

The majority of this manuscript was completed more than fifteen years ago. After our children were born, I lost the drive necessary to complete the project, and it lay dormant on successive computers. The celebration of the sesquicentennial of the Civil War, and in particular the 150th anniversary of this campaign, spurred me toward the finish line. As with all such endeavors, the author is deeply indebted to the many persons who helped with the project. Unfortunately, so much time has passed since the initial research was undertaken that many of those who rendered assistance may no longer be working for the listed institutions. Nonetheless, I am grateful to the individuals and repositories that provided the research materials necessary for the construction and completion of this project, and apologize in advance for any omissions. Aid in obtaining copies of primary documents was provided by Mike Meier, National Archives; Karen Kearns, H. E. Huntington Library; Glenn Longacre, Ohio Historical Society; John Geddis, Sally Childs-Helton and Laura J. Dragash, Indiana Historical

Society; Charlotte Ray and Gail Miller DeLoach, Georgia Department of Archives and History; Joseph Robertson, Elizabeth Gunshee, Janet Schwarz and Frances J. Pollard, Virginia Historical Society; Chris Kolbe, Library of Virginia; John L. Ferguson, Arkansas History Commission; Jaime Simmons, Harold Newman, Greg Carroll and Debra Basham, West Virginia Archives and History; Christelle Venham, Dave Bartlett and Danny Williams, West Virginia and Regional History Collection; the late Dr. Kenneth Slack and Jack Dickinson, Rosanna Blake Collection, Marshall University; Lisle Brown, Cora Teel, Kathleen Bledsoe and Andy Earles, Special Collections Department, Marshall University; Juanita G. Grant, Mary B. Blount Library, Averett College; Heather R. Munro and Becky Cape, Lilly Library, Indiana University; Beverly B. Allen, Robert W. Woodruff Library, Emory University; J. Larry Gulley and Melissa Bush, Hargrett Rare Book and Manuscript Library, University of Georgia; C. Vaughan Stanley, James G. Leyburn Library, Washington and Lee University; Gregory Johnson, Alderman Library, University of Virginia; the staff of the Preston Library, Virginia Military Institute; Mary Lou Pratt, Cabell County Public Library; and Pat Ward, Phillips County Library, Helena, Arkansas. A special thanks to Hunter Lesser, who generously provided photographs and research materials from his personal collection. Others providing encouragement, research assistance and/or copies of documents from personal collections were Amanda K. Adams, Richard L. Armstrong, Jack L. Dickinson, Tim McKinney, Bill McNeel, Leon and Bradley Nelson, Bill Patton, John Proctor, Becky Reger, Mike Sheets, Marc Tissenbaum, Tim Walton and Eddie Woodward. A special thanks to Robert Sawrey, the late Robert Maddox, Frank Riddel and Charles Bias for reviewing a master's thesis on this topic. I am extremely indebted to Debra Basham, Terry Lowry and Mary Johnson for undertaking the painful process of reviewing the manuscript. Terry's expertise in West Virginia's Civil War history was invaluable, while Mary provided plenty of red ink that made the finished product a bit more readable. Although she is admittedly not a military historian, she is an editor without equal, and I am thankful for her patience and perseverance in going through several drafts. All three are true friends who volunteered their evenings and weekends to edit this work.

I am grateful to my parents, Joe Geiger and June Geiger, and my step-parents, Marsha Geiger and the late John Snyder, for their unwavering support over the years. I am also thankful for my two wonderful children, Matthew Quinn (who got his middle name from his grandfather) and Rachel Elizabeth, and my best friend and wife Lois, who has put up with me for more than half our lives. I have undeservedly been truly blessed by a God who is always there for me in spite of my many shortcomings.

This book is dedicated to the soldiers from both sides who received their military baptism in 1861-62 along the Staunton-Parkersburg Turnpike in the mountains of western Virginia.

INTRODUCTION

Almost forty years after the close of the Civil War, Ambrose Bierce, one of America's literary giants, returned to visit the campsites and battlefields of West Virginia which were the basis for several of his short stories. Bierce had served with the 9th Indiana Infantry and spent most of the war's first year in the mountains along the Staunton-Parkersburg Turnpike. His last stop was Camp Allegheny, the site of the Battle of Allegheny Mountain, fought on December 13, 1861:

> *Here the regiment had its hardest fight in West Virginia, and was most gloriously thrashed. When I saw the place (with better opportunities for observation than we had then) I knew why. The works are skillfully constructed and nearly a mile in length, with placements for several batteries. They are built on a narrow ridge and are hardly more than one hundred and fifty yards wide at any point. At the rear, where our attack was made (after the garrison, having defeated our co-operating force in front, had gotten "good and ready" for us to surprise them) there was but one approach, and that was by a narrow road, through acres of slashed timber, impenetrable to a cat. The trunks of the trees are still there, all pointing away from the fort, all decaying, and none of them having even their largest branches. A big headlog across the embrasure commanding the road is so rotten that one can pick it to pieces with the fingers. I fancy the Yankee bullets have all been picked out of it; I found none.*
>
> *The slashed timber, which prevented us from attacking in line, saved our lives—most of them—when we attacked in column. We took cover in it and pot-shotted the fellows behind the parapet all day, as I recollect it, and then withdrew and began our long retreat in a frame of mind that would have done credit to an imp of Satan.*
>
> *The road that penetrated the slashed timber is easily traced; I recognized the spot where Captain Madden fell, at the extreme*

Postcard view of portion of Camp Allegheny. West Virginia State Archives

head of the column. Lord! how close to the works it was—I had thought it farther away.[1]

More than one hundred years after Bierce's visit, the battlefield on the summit of Allegheny Mountain has changed little. Chimney piles from cabins occupied by Confederate soldiers dot the landscape and the trenches they constructed can easily be traced for hundreds of yards. The old turnpike meanders between the remains of Camp Allegheny, providing a rustic reminder of the conflict waged across the United States. This ground, hallowed by the blood of soldiers from both North and South, has been remarkably preserved. Due in large part to its remote location, the Camp Allegheny site is one of the best places in the country to visualize the hardships faced by soldiers on both sides and to recapture the history which unfolded here. A little more than two weeks after the Battle of Allegheny Mountain, Charles Delay of the 25th Virginia wrote to his parents, "Our fortifications add very much to the safety and appearance of the place, it would be worth the trip for them [friends] to come out here to see our improvements." Delay's observation about the visual attractiveness of the site remains true. Twenty-five miles away, visible on a clear day, is Cheat Summit, which was occupied by Union troops in July 1861 and would serve as the winter home for thousands of men from Indiana, Ohio and Pennsylvania. The remains of the extensive fortifications they constructed are still evident today. Parts of both encampment sites, which were the highest campsites east of the Mississippi River, are preserved and interpreted by Monongahela National Forest.[2]

For most of the soldiers, it was their first time away from their native states. They found themselves fighting, suffering and dying in a bitterly daunting, yet hauntingly beautiful land hundreds of miles from their homes. Life was often difficult for the men who were stationed atop Allegheny Summit and Cheat Summit during the winter of 1861-62, and few enjoyed the environment. "There is more weather here than at any place I ever was at," Delay wrote. "It is raining or snowing almost every day at least four or five days in a week. Some times the mud will be shoe-top deep, and in less than twenty-four hours it will be frozen twelve or fifteen inches deep, so sudden is and has been the change of the weather on the top of the Allegheny Mountain." A Union soldier described the Cheat Mountain region as “a gloomy and melancholy place. We can see nothing but lofty mountains and snow-clad pines, which look like the monuments of bygone centuries.”

Another wrote, “I think we can get along in most any country except Old Virginy and that blessed state is calculated to kill any body that is half human, but I can stand it if any body can but I don’t care anything about trying it. If I was ever tired of any place in this world it is this God forsaken state. I would be willing to start at any time and not stop until I got ‘Out of the wilderness.’ I used to think I would like to live in a hilly country but that notion is about ‘played out’ in my mind as far as this part of the world.”

While many of the soldiers on either side despised the inclement weather and rugged terrain of western Virginia, others were transfixed by the beauty of the country. An Indiana soldier wrote: “I never gaze upon these old cloud-capped mountains, covered with their variegated foliage and presenting to the beholders so many beauteous scenes, but that I am made to love the land I would otherwise curse . . . the magnificent mountain scenery and crystal streams of Western Virginia reign without a rival. They are the mountaineer’s ideal and the soldier’s solace.”[3]

The Confederate soldiers who spent the winter of 1861-62 at Camp Allegheny came from Georgia, Arkansas, Kentucky, and Virginia. Many of them viewed the Civil War as a second war for independence. One wrote, “our Revolutionary fathers fought for seven long years for their independance and the independance that we once enjoyed; wee may have it to do.” Charles Delay hoped that “by the assistance of the ‘God of battles’, we may be able to drive the invaders from our soil our land of liberty, and that we may once more be a free and independent people.” According to Gilmer County native Christian Kuhl, who served with the 31st Virginia Infantry, “the fact that we were fighting for the cause we espoused and our families and loved ones, our homes and property, was a double reason why we gained the reputation our regiment enjoyed.”[4]

The Union soldiers came from Ohio, Indiana, Pennsylvania, and western Virginia and fought to re-unite the country. Charles E. Smith of the 32nd Ohio wrote that “the feelings which came over me, when I left home, friends and all that seemed dear, I cannot describe. I knew that I was facing danger and laying myself liable to be killed by the Rebels. I knew that I should have to undergo many hardships and be deprived of many privileges. But when I saw our great government, founded as it was by our noble ancestors, and our beautiful country, our free institutions under which we have lived, prospered and enjoyed the blessing of freedom and of God. When I saw it about to be torn down and destroyed, I felt it my duty to enlist in the cause of my country."[5]

This book seeks to provide a detailed look at military activities along the Staunton-Parkersburg Turnpike from mid-September 1861 to the first week of April 1862. This campaign, fought primarily in Pocahontas County, Virginia, included the Battle of Greenbrier River, in which nearly 7,000 soldiers clashed in what was primarily an artillery duel; the Battle of Allegheny Mountain, the bloodiest battle of the first year of the war in present-day West Virginia; the January 1862 raid on Huntersville; and numerous other skirmishes, raids, expeditions, incidents and events.

The evidence shows that although Union forces never planned an offensive eastward along the Staunton-Parkersburg Turnpike after Major General George B. McClellan departed from western Virginia, Confederate leaders were convinced that failure to defend this road would result in a Union advance toward Staunton, a belief that doomed hundreds of Confederate soldiers to spend a winter in a most inhospitable land.

The soldiers who lived through these tumultuous times would remember their experiences in this remote region for the rest of their lives, and this endeavor was undertaken and completed so that their sacrifices and experiences are documented and preserved for future generations. They would undoubtedly be pleased to be remembered.

CHAPTER ONE
WAR ERUPTS IN WESTERN VIRGINIA

"Soldiers: You are ordered to cross the frontier and enter upon the soil of Virginia," Major General George B. McClellan, commanding the Department of Ohio, wrote to his Union troops on May 26, 1861. "Your mission is to restore peace and confidence, to protect the majesty of the law, and to rescue our brethren from the grasp of armed traitors." On April 17, Virginia delegates had passed an Ordinance of Secession severing their ties with the United States, but the voters of the state would not vote on its ratification for another month. In the meantime, Virginia authorities proceeded as if secession was a foregone conclusion, mobilizing the state militia, seizing federal installations at Harpers Ferry and Norfolk, and forming an alliance with the Confederate States of America. Throughout western Virginia, town meetings were held both for and against secession, and thousands formed into local companies that would serve North and South. On May 23, Virginia voters approved the Ordinance of Secession, a decision that set in motion events which would eventually lead to the creation of a new state, West Virginia.[6]

Western Virginia was important to both North and South for several reasons. A primary consideration was control of the two rail lines in Virginia's western region, particularly the vital section of the Baltimore & Ohio Railroad running from Harpers Ferry through Grafton to Benwood, just south of Wheeling. The B & O provided the most direct route to Washington from the West; without control of the railroad, Union supplies and troops would have to be transported to the Federal capital in a roundabout manner. A branch of this rail line, the Northwestern Virginia, ran from Grafton to Parkersburg. Western Virginia was also important because of its abundance of livestock, farm crops, and salt; the latter was particularly scarce in the Southern states. Both sides also hoped to attract the manpower of the region to the services of their armies.

Early in the war, Major General George B. McClellan, who would command the Union force which invaded western Virginia, proposed to undertake an offensive along the James River and Kanawha Turnpike with

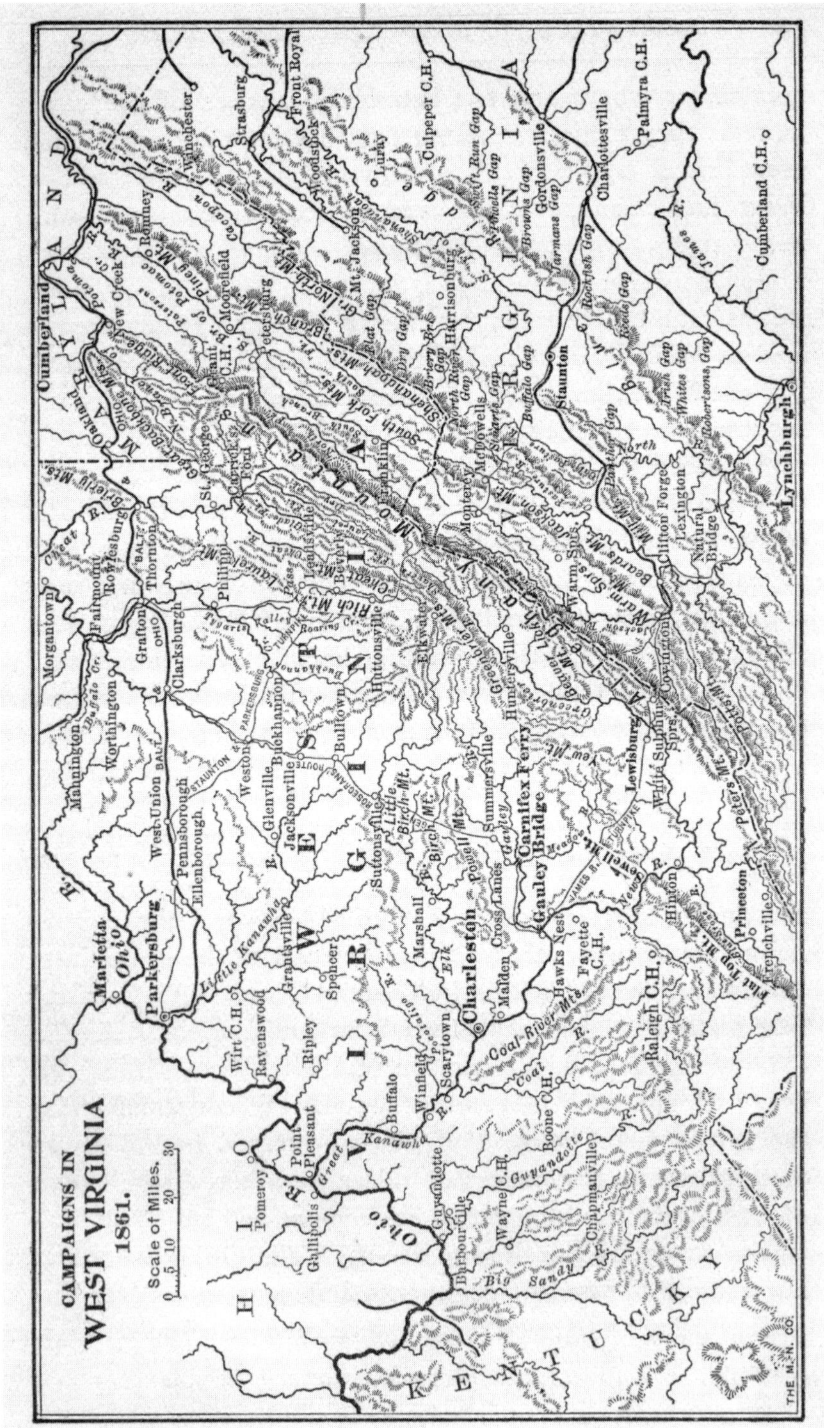

Western Virginia in 1861. Jacob Dolson Cox, Military Reminiscences (New York: Charles Scribner & Sons, 1900)

Union General George B. McClellan, whose success in western Virginia led to his promotion to head of the Union army. Library of Congress, Prints & Photographs Division, Civil War Photographs [LC-DIG-cwpb-07035 DLC]

the Confederate capital of Richmond as his destination. Although the plan was quickly judged unfeasible by Lieutenant General Winfield Scott, the general-in-chief of Union armed forces, he did not dispute the importance of military control of the region. According to author George E. Moore, "Had the Ohio River become a defensible frontier, the Virginia Panhandle could have been a base for military operations against communications between Washington and the Middle West." While Confederate leaders also recognized the importance of maintaining control of western Virginia, they initially failed to comprehend that a substantial military force would have to be imported to repel the inevitable invasion. Perhaps more important were the political considerations underlying the desire to control western Virginia. For the Union, military control of the region would be a huge triumph, offering protection to Union supporters from northwestern Virginia. For the Confederacy, western Virginia must remain part of Virginia, demonstrating a unified front.[7]

Confederate authorities apparently believed that a majority of the citizens of western Virginia would rise in arms to support the decision of the state's leaders to break from the Union, speculation which proved erroneous. The head of military forces in Virginia, General Robert E. Lee, sent dispatches to several respected military men of the region, who were ordered to form regiments to fight for the Confederacy. Colonel George A. Porterfield would command Confederate forces in the region with the mission to retain control of "both branches of the railroad to the Ohio River, to prevent its being used to the injury of the State." Porterfield's orders directed him to select a position near Grafton, where he arrived on May 14.

Prior to Porterfield's arrival, Major Francis M. Boykin, Jr., who commanded a small Confederate force which had gathered at Grafton, informed Lee that troops would have to be sent from eastern Virginia to hold the rail line. Boykin had discovered strong support for the Union in western Virginia, and his attempts to raise troops for the Confederacy were discouraged by Northern supporters, who expected imminent relief in the form of Pennsylvania and Ohio troops. Boykin noted that the western part of Virginia "is verging on a state of actual rebellion," an ironic statement considering the recent secession of Virginia.[8]

While Union supporters from northwestern Virginia repudiated the act of secession and met in Wheeling to decide upon a course of action, Federal troops from Ohio and Indiana crossed the Ohio River and quickly moved to clear the region of Confederate troops. Porterfield's small Confederate force, which had withdrawn to Philippi, was routed there on June 3 in the first land battle of the Civil War. Belatedly recognizing the looming disaster in the region, Confederate military officials sent reinforcements under the command of Brigadier General Robert S. Garnett to check the Federal advance. Garnett fortified positions at Laurel Hill and along the western slope of Rich Mountain, but a Union force outflanked him and defeated Confederates stationed along the Staunton-Parkersburg Turnpike at Rich Mountain on July 11.

Erroneously believing that McClellan's army occupied Beverly, thus cutting off his line of retreat along the turnpike, Garnett ordered a northeastward withdrawal which quickly turned into a rout. General Garnett was killed at Corrick's Ford and his army was driven out of western Virginia. Less than two months after McClellan's army entered western Virginia, the Staunton-Parkersburg Turnpike west of the Alleghenies had been cleared of Confederate forces.[9]

Prior to the Battle of Rich Mountain, McClellan had proposed to march along the pike to Staunton in the Valley of Virginia, a plan that initially alarmed Winfield Scott, who feared that the supply line required for such an advance would be too lengthy "without intermediate supports." Following the decisive Union victory at Rich Mountain, McClellan confidently wrote to Scott, "Will a movement of mine on Staunton facilitate your plans? If so, I can probably take that position. I do not know your plans of operation, but can move on to Staunton if you desire." Scott replied that McClellan's "suggestions in respect would be admirable." At the time, the Union Army of Northeastern Virginia was preparing to assault the Confederate Army of the Potomac near Manassas in eastern Virginia, not far from Washington, DC. If the attack was successful, and McClellan advanced to Staunton, the Union forces perhaps could link up near Winchester.[10]

The Battle of Manassas, fought on July 22, was a Union disaster, leading McClellan to change his opinion regarding an advance on Staunton. Shortly after receiving word of the Confederate victory, McClellan wrote, "My three-months' men are homesick and discontented with their officers, and determined to return at once. When I suggested the Staunton movement I expected these regiments to unite in it. I should be compelled to fight the enemy now ascertained in force at Monterey, and should reach Staunton without men enough to accomplish much." McClellan instead proposed to move by rail to join forces with Generals Robert Patterson and Irvin McDowell in order to contend with the victorious Confederates threatening Washington. He concluded that this movement, "though not so brilliant a plan as a movement on Staunton, appears to me the sounder and safer one."

Scott agreed the plan to advance on Staunton was no longer feasible and on July 22 ordered McClellan to "remain in your present command instead of going to the Valley of the Shenandoah." Later that day, McClellan received orders directing him to Washington, where he assumed command of the Union Army of the Potomac; he was replaced in western Virginia by Brigadier General William S. Rosecrans.[11]

Union military leaders were content with fortifying and occupying the mountain passes to prevent any Confederate attempt to regain control of the region. On August 4 General George B. McClellan wrote President Lincoln, "all the passes into Western Virginia from the East should be securely guarded, but I would advise no movement from that quarter towards Richmond, unless the political condition of Kentucky renders it impossible or inexpedient for us to make the movement upon Eastern Tennessee through that State."[12]

McClellan's failure to press the offensive along the Staunton-Parkersburg Turnpike over the Alleghenies allowed Confederate military leaders to reorganize the scattered remnants of the Confederate army in western Virginia and dispatch more reinforcements from throughout the Confederacy. Brigadier General William W. Loring, who was placed in command of the Confederate Army of the Northwest on July 20, was informed by General Robert E. Lee of the "necessity of preventing the advance of the enemy, and the importance of restraining him the other side of the Alleghany Ridge." Loring was directed to occupy passes for that purpose, particularly "the defense of the road leading from Huttonsville... to Huntersville, and the Warm Springs to Millborough, on the Virginia Central Railroad."

On the following day, Lee wrote Loring that "the selection of those passes is, of course, left to your judgment," but he recommended that Cheat Mountain, on the Staunton-Parkersburg Turnpike, and Middle Mountain,

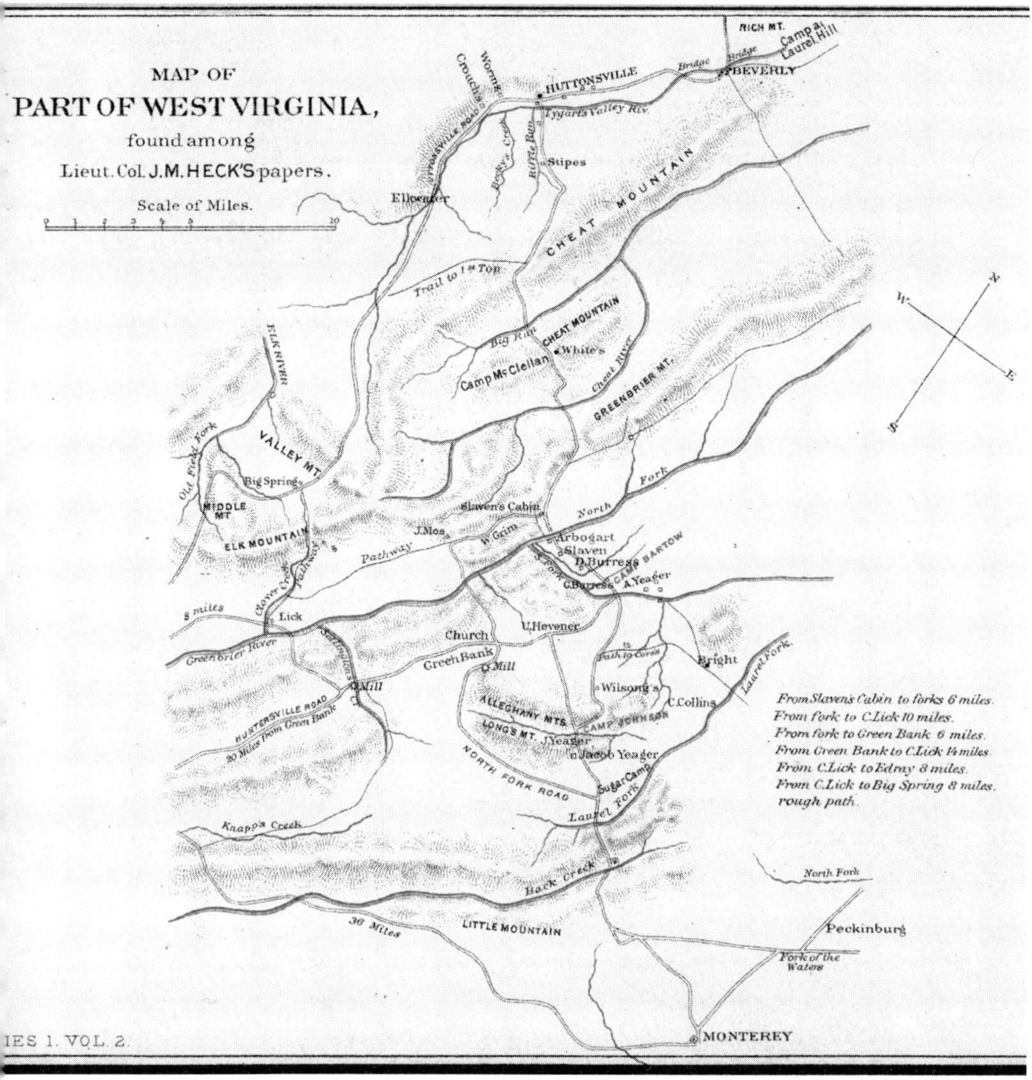

Map of area of operations, parts of Pocahontas and Randolph counties. Official Records

on the Huttonsville-Huntersville Turnpike, be occupied if McClellan had not already taken positions at those points.[13]

Loring moved part of his army to Middle Mountain per Lee's suggestion, but Cheat Mountain was already occupied by Federal troops. On July 28, Loring wrote, "There is said to be a strong position in advance of the Alleghany Pass, at Yeager's, near the crossing of the Greenbrier River." Brigadier General Henry R. Jackson soon advanced along the Staunton-Parkersburg Turnpike from the summit of Allegheny Mountain to the Pocahontas County location noted by Loring. This site not only controlled the turnpike, but was also in close proximity to a north/south road leading through Green Bank and Huntersville to the Virginia Central Railroad at Millboro. Located here was Traveler's Repose, the first stagecoach stop west of the Alleghenies. The home, owned by Andrew Yeager, also served as an inn and post office. The camp constructed here became known as Camp Bartow, named in honor of a Confederate officer from Georgia who was killed at the Battle of Manassas.[14]

General Robert E. Lee, who was ordered to the region as a military advisor following the death of General Robert S. Garnett, would eventually gather enough troops to plan a counteroffensive against the Union army in the Cheat Mountain region that was designed to dislodge Federal forces from positions at Elkwater, Huttonsville, and along Cheat Mountain.

Lee arrived in Huntersville on August 3 and three days later located his headquarters at Valley Mountain on the Huttonsville-Huntersville Turnpike. Four Confederate brigades, designated the Huntersville Division of the Army of the Northwest, were stationed here and at other points defending this road. The Monterey Division, comprised of two brigades, occupied Camp Bartow.[15]

General Lee spent countless hours reconnoitering the Union defenses in the area. The fortifications at Elkwater and Cheat Summit were impressive, and he realized a frontal attack upon either would be disastrous. Lee eventually discovered trails that led to the rear of Cheat Summit Fort, and he decided to outflank both Union positions in a simultaneous attack. Despite the early loss of western Virginia, Confederate leaders were hopeful the situation could be reversed when Lee's attack was initiated. On September 5, Confederate President Jefferson Davis wrote that, had it not been for the torrential rains, "I think he would have beaten the enemy in that quarter. Had we the means to

Confederate General Robert E. Lee. Library of Congress, Prints & Photographs Division, Civil War Photographs [LC-DIG-cwpb-04402]

move on Beverly from Winchester, it might result in the capture of Rosecrans and the repossession of Western Virginia."[16]

On September 9, the day the offensive was launched, General Lee wrote to the Confederate soldiers, "The eyes of the country are upon you. The safety of your homes and the lives of all you hold dear depend upon your courage and exertions. Let each man resolve to be victorious, and that the right of self-government, liberty, and peace shall in him find a defender. The progress of this army must be forward."

The offensive called for six brigades to move into position; the opening of an attack from Colonel Albert Rust's brigade was to be the signal for the coordinated assault to begin. The Confederate brigades reached their positions, but Rust failed to attack because he believed his force to be outnumbered. Lee and Loring advanced on Elkwater on September 12 but were repulsed by the Union defenders.

Lee's well-laid plans went for naught due to Rust's failure to launch an attack on Cheat Summit Fort. The complicated Confederate assault on Cheat Mountain in September 1861 was well-conceived but unsuccessful due to the timidity of some of Lee's subordinates. Following the failure, no further efforts were made by Confederate forces to advance westward along the Staunton-Parkersburg Turnpike.[17]

In addition to the disasters along the turnpike, the Confederates had been forced from the Kanawha Valley in July. Despite repelling repeated Union attacks at Carnifex Ferry on September 10, the Confederate Army of the Kanawha had been compelled to withdraw from the battlefield due to the overwhelming numerical advantage possessed by the Federals, thus ending a Confederate attempt to retake the Kanawha Valley.

After the defeat at Cheat Mountain, Robert E. Lee, accompanied by a substantial portion of General William W. Loring's Army of the Northwest, linked up with Brigadier General John B. Floyd at Big Sewell Mountain where they attempted to lure the Union force commanded by General William S. Rosecrans into battle. Rosecrans, however, withdrew in early October to the Federal supply base at Gauley Bridge. Following the failures at Cheat Mountain and Big Sewell Mountain, Confederate military leaders exerted little effort to regain control of western Virginia during the first year of the war and sought primarily to prevent a Union advance over the Allegheny Mountains. They continued to believe that a Federal offensive along the Staunton-Parkersburg Turnpike was imminent.[18]

The Staunton-Parkersburg Turnpike connected Staunton in the Valley of Virginia with Parkersburg, located along the Ohio River. Authorized in 1838 and completed within a decade, the road led to increased settlement and commerce in western Virginia. During the Civil War, its importance was magnified because of the scarcity of good roads in western Virginia. Union control of the turnpike following the initial campaign ensured protection for the B & O Railroad, and their advanced positions at Cheat Mountain presented the possibility of a forward movement toward Staunton and the vital Virginia Central Railroad, one of the most important rail lines in the Confederacy.

In the aftermath of their meeting along Cheat Mountain, Union and Confederate forces in the region prepared for the next move in the campaign. Federal units, stationed in Randolph County at Elkwater, Huttonsville and Cheat Summit, included the 7th, 9th, 13th, 14th, 15th and 17th Indiana Infantry regiments; the 3rd, 6th, 24th, 25th and 32nd Ohio Infantry; and the 2nd (West) Virginia. Union forces comprising the Cheat Mountain District of the Army of Occupation, Western Virginia, were commanded by Brigadier General Joseph Jones Reynolds. A Kentuckian by birth, at age 15 Reynolds had moved across the Ohio River with his family to Lafayette, Indiana. Reynolds graduated from West Point, class of 1843, and a few years later began teaching at his alma mater, where he remained for eight years. After resigning from the military in 1857, he taught engineering at Washington University in St. Louis. At the outset of the war, Reynolds was appointed colonel of a militia regiment, but he was quickly promoted to brigadier general of United States Volunteers. His aide-de-camp, Lieutenant Isaiah B. McDonald, wrote Governor Oliver P. Morton, "The interests of the country are safe in the hands of Gen'l. Reynolds, and I know he will look well after the fame of old 'Hoosier State.' All he wants is more men."[19]

Two brigades of the Confederate Army of the Northwest, led by General Henry Rootes Jackson, were stationed to the east at Camp Bartow and nearby Allegheny Summit in Pocahontas County, and included the 1st and 12th Georgia Infantry regiments; the 23rd, 31st, 37th, 44th, and 52nd Virginia Infantry regiments; the 9th and 25th Virginia Infantry battalions; and the 3rd Arkansas Infantry. Jackson, a Georgia native, graduated from Yale University in 1839 and, after practicing law in Savannah for a few years, was appointed a United States district attorney at the age of 24. During the Mexican War,

Jackson served as colonel of a Georgia volunteer regiment. Later, he was a newspaper editor, superior court judge, and United States minister resident to Austria, before serving as a member of the Georgia secession convention. After the decision to break with the Union, he was appointed brigadier general in the Provisional Confederate Army on June 1, 1861, and soon found himself embroiled in the campaign in western Virginia.[20]

The region where these two armies camped and clashed was extremely mountainous and sparsely populated. Randolph County had slightly fewer than 5,000 residents in 1860, while the population of Pocahontas County was just less than 4,000. The majority of citizens in both counties supported the Confederacy. Most were farmers, and many had never ventured far from their homes. One Indiana soldier scouting in the Cheat Mountain region stumbled across an old log cabin occupied by an elderly man and his wife "who must have been at least one hundred years old." The man had moved to Randolph County seventy-five years earlier and had not left the county since, and he informed the Union soldier he had not seen an American flag since he was a boy. After learning he could see one flying at the nearby Union camp, the elderly man "hobbled out and got on a horse a[nd] started of[f] as though he was going after a fortune."

Pocahontas County hosted the headwaters of a number of waterways, including the Greenbrier River, a tributary of the New River formed by two forks in the northern part of the county. Over the next few months, numerous skirmishes would be fought near this confluence.[21]

It was in this mountainous region that most of the soldiers under Reynolds and Jackson were destined to spend a bitter winter. "I don't see what a country like this was ever intended for," wrote one Indianan about his home away from home, his words undoubtedly reflecting the thoughts of many of the soldiers wearing blue or gray. "I have traveled for ten miles along the pike and not meet [sic] a living person. White houses and farms are as scarce as angels visits to a soldiers camp, it 30 miles in a direct line east from our camp before you come to a cleared place or a house not quite as thickly settled as some parts of this world that I have been in but as I said before this country was never made for a human to live in so I have no right to grumble."

Fierce rain storms would suddenly appear in the region, and as a correspondent to the *Cincinnati Daily Commercial* informed the paper's

General Henry R. Jackson, who commanded the Confederate force stationed at Camp Bartow. Courtesy of Hunter Lesser

readers, "when the dark clouds thunder stored [sic] gather upon their brows, the scene is terrible. Rain descends in a solid body, and the tiny stream, with a 'presto-change,' is converted into a mighty river that dashes madly along, carrying all things before it with a force equal to that of the cataract of Niagara."[22]

Though many of the Federal soldiers despised the harsh conditions, at times their admiration for the beauty around them could not be suppressed. Private John Hadley and two other men from the 7th Indiana climbed the highest peak above their camp at Cheat Mountain Pass. Writing to his brother, Hadley remarked, "I shall not attempt to describe to you the sublimity of the scenery—I should fail even if I were able to command every strong ajective in the English language. A single narrow vally with occasionally a house dotting over it, was all that could be seen except mountains mountains! Mountains!! Mountains!!! I never expected to see the like, neither did I ever dream that such existed. I seemed to see ten thousand mountains, one rising a little above the other until the highest towered far above the clouds."

Another soldier found the region "Beautiful, most beautiful." Contending that only "half its beauties have yet been sung," he described "vallies verdant and fresh as those trod by our first parents," surrounded by mountains, "Olympus-like and cloud-capped, with their sides clothed with foliage of every hue, tine and texture," and from which cold streams ran, bringing water to the valleys. "Viewed from some distant sister peak, the undulating piles look smooth as some close-shaven lawn," he continued. "Its mighty elements lie concealed from the casual glance, and truly the road to death is through pleasant places."[23]

It was difficult for the Federal government to supply its army in this beautiful, yet fierce region. The only rail lines were the B & O, which passed through Grafton and terminated in Wheeling, and the Northwestern Virginia, which connected Grafton and Parkersburg. Union supplies were sent in wagons from Grafton to Cheat Summit, a distance of about seventy miles. Confederate supplies were also transported along the turnpike westward from Staunton. The Staunton-Parkersburg Turnpike had been built in the 1830s and was a well-built road for its time, but the wear caused by the continual use and the seemingly ever-present rain, which turned the pike into a muddy quagmire, soon made it difficult to traverse this route.

Although supplying the Federal troops in the Cheat Mountain region proved difficult, some of the soldiers ate well. At times the men dined on fresh beef, including fried beefsteak, mutton, stewed peaches, bread and hard crackers. Potatoes were dried and ground up like hominy. The surrounding area provided wild grapes and chestnuts galore. Cheese could be purchased for five cents a pinch, tobacco was ten cents a chew, and whiskey was one dollar a pint. However, describing the fare for six of the seven days in a week, Frank Ingersoll of the 13th Indiana wrote, "four days in the week we fast and the other two days we live on pies and turkey, baked beans, dessicated vegetables." This latter delicacy was comprised of "equal parts of beet leaves, dog-fennell carrots, various stems, potatoe vines, and about 700 kinds of weeds, all steamed and presed up together so that a piece an inch thick will make soup enough for 50 men."[24]

Supplying the soldiers stationed at Camp Bartow also proved a difficult task for Confederate authorities. Thousands of bags of grain had been sent to his command, but Quartermaster Michael G. Harman was forced to impress wagons and assure local farmers their teams would only be used for one trip. He was having a difficult time finding horses and wagons to haul provisions to Camp Bartow, but Harman pledged to supply Jackson's army with corn and oats for the horses if the wagons and empty feed bags from previous trips were returned in a timely manner.[25]

Several of the regiments at Camp Bartow received orders from General Robert E. Lee to prepare to march at noon on September 15, but they did not leave camp until 4 p.m. The troops supposed they were either heading to Big Spring to join General Loring's command or marching to the relief of General John Floyd, whose command was rumored to be surrounded in Greenbrier County. Arriving at Green Bank about 9 p.m., the men laid down for the night without pitching their tents.

In the early morning hours of September 16, rain began to fall, awakening the troops from their blissful slumber. A second Confederate force moved west along the turnpike and bivouacked at Slaven's Cabin, where they found muskets, blankets and other equipment abandoned by Union pickets when Lieutenant Samuel Dawson's men "lit on them" during the attack on Cheat Summit. Five of these men, led by Sergeant James Atkins, made their way through thick underbrush and fallen trees to reconnoiter the Union camp. Their efforts were exerting but uneventful.[26]

The 31st Virginia remained at Green Bank until 9 a.m. on September 16, when they were ordered to return to Camp Bartow. The road was extremely muddy and the rain returned in the evening. Spirits were low in the Confederate camp. Private Joseph C. Snider, who was still fighting the after-effects of the measles, was cheered by "a good mess of ripe apples," but his elation was only temporary. That evening Snider wrote: "I feel more despondent to day than I have since I have been in the service but still trust in the justice of our cause and the intervention of an allwise providence." His frustration was echoed by many of his fellow Confederates who were angered by the botched attack on the Union positions. The force that had once again approached Cheat Summit began preparing fortifications when they received orders to return to Camp Bartow. Many of the soldiers fell prey to illness caused by the hardships of the expedition. Nearly all were discouraged by the failed offensive which accomplished little and "has only caused great suffering."[27]

September 17 was another rainy day. While on picket duty, Private William Holt of the Barbour Greys somehow shot his own left hand, and a surgeon was forced to amputate two of his fingers. Later in the day Captain Lindsey M. Shumaker, who commanded an artillery company, became embroiled in an argument with Captain Felix Hull of the 31st Virginia. Shumaker became so incensed that he fired a shot at Hull but, fortunately for both, he missed.

Fearing that a Federal force would advance from Petersburg (in present-day Grant County) south to Monterey, General Henry R. Jackson had stationed the 52nd Virginia at Forks of Waters, seven miles north of Monterey. The 3rd Arkansas and 37th Virginia, along with Captain Shumaker's artillery company, began withdrawing from Camp Bartow to join the 52nd that evening. Before reaching their destination, the order was countermanded due to Jackson's fear of an attack upon his command along the Greenbrier River, even though the possibility of turning the tables and threatening Petersburg was tempting. A Confederate offensive against the town would potentially permit disruption of the Baltimore and Ohio Railroad and also cut off communication with the Union forces at Cheat Mountain, possibly forcing a withdrawal from their impregnable works in the region. The troops returned to their camps; the 3rd Arkansas and Shumaker's artillery to Camp Bartow and the 37th Virginia to Green Bank.[28]

At Camp Bartow, now occupied by two brigades of the Army of Northwest Virginia, work on the fortifications progressed rapidly. Trenches were dug along the hillsides and trees were felled for half a mile in every direction. The timber was then piled up, forming an impenetrable blockade. The soldiers' confidence in the bulwarks was so high that the men actually hoped to be attacked. One Georgia soldier boasted, "In a few days this encampment will be so strongly fortified that the force now here could resist successfully an attack of 10,000 of the enemy; at any rate, we would like to receive a call from that number of our Yankee friends in our fortifications . . . we will ever be ready to welcome the ruthless invaders of our soil 'with bloody hands to hospitable graves'." In expectation of an attack, a "very large grand guard" was sent out for more than twenty-four hours to scout the vicinity.[29]

The Georgia soldiers in particular had no wish to remain in western Virginia during the harsh winter months. One hoped that reinforcements from the eastern part of Virginia would enable the Confederates to drive Union forces beyond the Ohio River. He noted, "we don't like the idea of having to quarter in these Mountains during the ensuing winter, which will necessarily be the case unless we succeed in expelling the Hessians from Western Virginia."

Nearly half of the Confederate soldiers at Camp Bartow were ill. Indicative of the widespread sickness is a Georgia sergeant's report that only 21 out of the 101 men in his company were able to report for duty. To guard against false claims and to stem the large number of soldiers who reported sick, a general order was issued declaring that those who reported "unfit for duty" were to be examined by a board of medical officers; men found not to be ill were to have charges preferred against them. The Confederate sick were shipped east to Crab Bottom on September 25 in anticipation of an attack.[30]

Despite the thinned ranks, small scouting parties combed the mountains to detect any movements by the enemy. On September 25, Confederate scouts stopped at the home of John Slayton in Pocahontas County, where they procured a bushel of apples and a gallon of honey. Slayton, wrote one Confederate, "has decamped and is supposed to be with the Yankees." His assumption was correct. John Slayton would prove a considerable asset for Federal forces in the region. He was 33 years old when the war broke out and was determined to aid the Union despite the danger he faced residing in a county with strong sentiment in favor of the Confederacy.[31]

"There isent much prospects of a fight here now," Shep Pryor of the 12th Georgia wrote his wife on September 24. "At the same time they are diging out ditches all around our camp, cutting down all the timber for a half mile around the camp, falling them all [the] same way so that it is impossable to pass through them. Our officers must be expecting an attack here."Their expectations would be fulfilled shortly.[32]

CHAPTER TWO
A PLAN TO PAY RESPECTS

Shortly after the September 10, 1861, Battle of Carnifex Ferry, General William Starke Rosecrans, commander of the Union forces in western Virginia, had warned Union General Joseph J. Reynolds that his command in the Cheat Mountain region was entrusted to hold the enemy in Pocahontas County: "The rebels have retreated over Big Sewell, if not to Meadow Mountain. They may join forces and try to crush us with their augmented strength. They cannot crush you."

Rosecrans had ordered Reynolds to maintain control of the turnpike far in advance of his fixed positions, which he was to hold securely. He also had encouraged Reynolds to advance against the Confederates, writing, "the moment you see any signs of an opportunity, fall on them, and worry and harass them if you can do no more. If your troops move lightly, carrying the necessary provisions to go and return, you can so harass them that you may prevent their leaving their position and stop any force they can bring against you. Hasten up your troops, and when your column is re-enforced we may have the opportunity to strike a decisive blow." Reynolds must have responded positively to the suggestion to take the offensive since two days later Rosecrans wrote to General Jacob D. Cox, stating that Reynolds, "getting re-enforcements, says he will pay his respects to Lee soon."[33]

Reynolds began planning for an advance on the Confederates at Camp Bartow. Scouting parties were sent out daily, as on September 24, when a squad of the 9th Indiana, known as the Swamp Devils, brought in five horses, fifty head of cattle, and three or four prisoners. None of the captured men were in uniform; one was a boy, 12 to 13 years of age.

On the next day, the men of the 15th Indiana took target practice in the morning, received marching orders and arrived at Cheat Summit at 9:00 p.m. Here they joined the 9th Indiana, the 14th Indiana and the 32nd Ohio. Two pieces of rifled cannon were also hauled up the mountain. The troops were ordered to prepare two days' rations; an advance on Camp Bartow would begin at midnight. Although the day was clear, a drenching rain began to fall about dark, and Reynolds postponed the offensive. It appears,

Union General William S. Rosecrans. Library of Congress, Prints & Photographs Division, Civil War Photographs [LC-DIG-cwpb-06052]

however, that the decision to cancel the attack was due not to weather conditions, but, instead, to Reynolds's determination to further increase his troop strength before undertaking an attack against the Confederate camp. Many of the Union soldiers were disappointed by this delay. "We all thought until an hour ago that we would make an onward movement tonight," one of them wrote, "but the order has been countermanded, and we now know nothing about what we will do. This is the 3rd time we have prepared to move, and have been disappointed."[34]

A September 26 dispatch from Reynolds to Colonel Nathan Kimball, in command of the 14th Indiana, adds further support for this contention. In it he stated that "by exercising a little patience we can attack them with an increased force of at least one Regiment," as well as ten pieces of artillery. Reynolds added, "now it seems to me this difference will more than compensate for what additional entrenchments they can make in two days—not doubting you could take them now think we had better wait the result of our reconnoissance towards Big Spring—so watch them close and be patient."[35]

Reynolds ordered the reconnaissance to determine the location and strength of the Confederate force stationed along the Huttonsville-Huntersville Turnpike. Before moving against Camp Bartow with a large force, Reynolds had to be certain there was not an enemy force along this route strong enough to seize the fortifications at Elkwater and Cheat Summit, which would be lightly defended once the advance by his army commenced.

The Union expeditionary force was headed by Colonel Jeremiah Sullivan and consisted of Sullivan's own regiment, the 13th Indiana; the 6th Ohio; two ten-pound rifled guns from Captain Cyrus Loomis's Michigan battery; and a detachment of Bracken's Cavalry.

Jeremiah Cutler Sullivan, a native of Madison, Indiana, was the son of attorney Jeremiah Sullivan, who served on the Indiana Supreme Court and gave Indianapolis its name. The younger Sullivan graduated from the United States Naval Academy and served six years in the Navy; however, when war erupted he formed an infantry regiment, the 6th Indiana, and he was later appointed colonel of the 13th Indiana. Sullivan's detachment, led by an experienced guide from the area, left Elkwater at 2 p.m. on Thursday, September 26, and marched southwest, covering ten miles the first day.

Union General Joseph J. Reynolds. Library of Congress, Prints & Photographs Division, Civil War Photographs [LC-DIG-ppmsca-22323]

Before nightfall, a hard, steady rain commenced, which would continue for days. As the troops were without tents, the rain made for lousy sleeping weather, although some of the lucky ones slept in an old saw mill five miles north of Marshall's Store. Strong pickets were ordered out, but the night passed without incident, and the column resumed the march at six on the morning of September 27.

Companies C and F of the 6th Ohio and Company C of the 13th Indiana were detailed to march a circuitous route of about six miles and begin the expected battle by striking the Confederates on their flank, which was supposed to be near Marshall's Store, owned by Jacob Marshall. (Marshall, who operated a grist mill here, served as a Confederate scout and guide and would soon organize and serve as captain for a company of the 19th Virginia Cavalry.) Although the maneuver was skillfully executed by the three companies, no Confederates were found at Marshall's. There were indications, however, that a detached camp which was discovered had only recently been deserted. Marshall's grist mill and store were burned, and Colonel Horace Heffren and Major Robert Foster of the 13th Indiana led Company B of the 6th Ohio and Company H of the 13th Indiana forward to probe for Confederate pickets.

The Confederate force was, in fact, at Big Spring, about three miles below Marshall's, and the Union force soon stumbled upon the Confederates' advance post about a mile north of the camp. They quickly drove the pickets in and advanced to a position where they could see part of the encampment, which was protected by ravines and earthworks. Though some estimated the camp contained as many as 7,000 Confederates, Sullivan reported there were closer to 1,500 enemy troops awaiting his advance.

A Confederate regiment appeared and opened fire, and a brief skirmish erupted. The Union advance was reinforced by five companies and an artillery piece, and Major Foster drove the Confederates back across the flats. Foster deemed it necessary to withdraw, however, not only because the Union force was outnumbered, but also because the torrential rainfall would soon cause flooding that would prevent their return. One Confederate soldier was captured; Union reports record no Federal casualties but several for the enemy. The prisoner revealed that General Lee had determined that, with no possibility of receiving supplies during the harsh mountain winter, the Confederates were to retire until the spring. He also asserted that sickness was especially prevalent.[36]

The heavy rains, which began on the evening of September 26, continued for the next three days, flooding the many rivers in the region and thwarting Reynolds's plan for a forward movement. At least two soldiers from the 15th Indiana and more than a dozen horses died from the cold and exposure. Two others were killed when a tree fell on their tent. Joseph Howell, the sutler for the 14th Indiana, drowned when the raging waters of the stream he was crossing swept him away. His mutilated body was discovered more than a mile downstream. The expeditionary force led by Sullivan had a difficult time returning to Cheat Mountain.

At one stream, the men were forced to cross on narrow logs or wade across the swollen waterway to reach the far side. Many of the soldiers were swept away and in danger of drowning, only to be grabbed and rescued by their comrades. Others clung to rocks until they could be saved. Francis Guhra, a native of Cincinnati who served in Company K of the 6th Ohio, slipped while crossing and was "whirled away with a rapidity of a shot."

When the men reached Elkwater, they found the camp had been submerged under four feet of water. All of the Union regiments in the Cheat Mountain region were exposed to the terrible storm, and many of the men had no overcoats or blankets. Some of the less fortunate were without shirts, pants, or socks.[37]

On Cheat Mountain, the rain and cold made sleeping a difficult task. The men spread pine branches on the ground in their tents and attempted to doze off. Steam from the soldiers' bodies "rose up from us all like clouds and as it cooled, came down like rain and wet our blankets."The fierce storm also caused discomfort for Confederates in the area. At Forks of Waters in Highland County, Virginia, where the 52nd Virginia was encamped, a soldier noted, "We had to move in a hurry, we had to wade through water over knee deep and we had to carry all of our things out about one hundred and fifty yards out on a hill and when we got all of the things carried out it was dark and then we had our tents to put up after dark. I tell you we had a wet time of it."[38]

The rain finally ceased on the morning of September 29. A heavy frost covered the ground in the morning, but the sun's rays soon warmed the ground and made for a beautiful Sabbath day. In the Union camp at Cheat

Summit, a newspaper correspondent witnessed a touching scene:

> *While sitting upon the fortifications, this morning, taking a distant view of the Alleghanies, the strains of a funeral dirge, from a full brass band, fell upon my ears. Looking to the opposite hill, I saw a funeral procession winding its way through Col. Ammon's camp. A young soldier, D. Summermire, of Company D, Twenty-fourth Ohio regiment, from Adams county, had died last night, and he was being borne to his grave on the shoulders of his comrades.*

The chaplain of the regiment held a lengthy service at the grave site, which was attended by several hundred troops. The ceremony was portrayed as "a most solemn and impressive sight."[39]

Sickness was taking a heavy toll on both Union and Confederate soldiers in the region. Private James Hall of the 31st Virginia bemoaned their fate, writing, "So common has death become, that when a man dies, he is as soon forgotten. Yesterday I passed by the graveyard of our Regt., it being in a line of the timber which we were felling as a blockade. A few tall hemlock pines were left around them, in respect for the dead. They lie far from the road, in a secluded spot. This may possibly be our fighting ground. The din and clangor of battle may sweep over them, as opposing squadrons meet in terrible combat. But they will sleep on."[40]

In times such as these, men tend to turn to religion as a comforting relief from the ever-present specter of death. On the afternoon of Summermire's funeral, a more uplifting service was held at the Union fort reflecting this emotion. The chaplain of the 9th Indiana preached to soldiers from a number of regiments, delivering his sermon from the upper parapet of a battery while his attentive audience occupied the lower breastworks. The chaplain's message was described as "an eloquent one, such, I am sure, as was never before preached in the mountains."

A rousing sermon was also given in the evening to the Confederate troops at Camp Bartow. The message on the 15th and 21st chapters in the book of Job, presented by the Reverend Robert Nelson Crooks, a first lieutenant in the 31st Virginia, must have been especially inspiring, since several of the soldiers noted the sermon in their diaries.[41]

Camp Bartow commanded both the turnpike which wound over the mountains to Staunton and the road leading through Green Bank south to Huntersville, about twenty miles distant. However, the Confederates under General Henry R. Jackson did not control a rough country road known as the Slaven's Cabin road, which branched off the turnpike about midway between Cheat Summit and Camp Bartow and terminated at Green Bank. By this road, a Union force could advance to Green Bank, where they could either attack the Confederates at Camp Bartow from the rear or march south through Huntersville to Monterey, cutting off the Confederate supply line from Staunton. Jackson had been instructed to hold his present position at Camp Bartow to prevent the enemy from advancing along the Slaven's Cabin road. Since the Confederate camp on the Greenbrier did not command this road, the only way Jackson could prevent a Union advance along this route was to attack the Federals at Cheat Summit, a task which had failed miserably only two weeks earlier. Jackson recognized the importance of the Slaven's Cabin road, but he felt that attempting to maintain a picket post at its juncture with the turnpike would be too costly due to the position's proximity to Federal scouting parties from Cheat Summit.

Jackson's cavalry had scouted daily along the road but had become so reduced in strength that he was unable to send mounted troops to reconnoiter. Jackson suggested that Major William Fitzhugh Henry Lee be instructed to send scouts out from his position at Clover Lick to warn of any Union flanking movement on the Slaven's Cabin road. He also requested reinforcements since he was required to guard not only the turnpike, but also the lines from Monterey, Franklin, and Petersburg.[42]

In a September 29 dispatch to Loring's adjutant, Colonel C. L. Stevenson, Jackson expressed his belief that Allegheny Summit was a far more effective defensive position than Camp Bartow, requiring fewer troops than the more vulnerable position presently held at its western base. Jackson also informed Stevenson that the efforts to construct defensive works at Camp Bartow required an immense amount of time and labor. Jackson's dilemma had been magnified by orders to also build fortifications on Allegheny Summit. He determined that this would be impossible without weakening his present position on the Greenbrier and, therefore, ordered Colonel John Baldwin of the 52nd Virginia Infantry to advance his regiment by the first of October from Forks of Waters to Allegheny Summit, where they would be placed in position by one of Jackson's staff officers. Before marching, Baldwin instructed Captain George Jackson of the cavalry to "keep a constant and vigilant eye" on the Petersburg, Franklin and Seneca roads to prevent a flanking movement which would lead to the rear of his force. Jackson would give immediate notice of any movement of Union forces in this region, which lay north of Camp Bartow.[43]

On the first day of October, Confederate General Henry R. Jackson sent another dispatch to Colonel C. L. Stevenson at the headquarters of General William W. Loring. Jackson once again informed Stevenson that his command was not strong enough to detach any part of it to the top of Allegheny, and he therefore had ordered the 52nd Virginia to this point to protect the rear of his army by commanding the road from Heavener's. He also noted that Lieutenant Colonel Seth Barton, who was overseeing the construction of the defensive works, believed that two or three more days of work would render Camp Bartow as strong as it could be made. Barton, a West Point graduate, had served for years in the western United States before resigning in June 1861 to support the Confederate cause.[44]

With winter rapidly approaching, plans were made for the upkeep of the roads in the region. Quartermaster-General A. C. Myers directed a dispatch to Captain C. R. Mason, assistant quartermaster, assigning him to superintend the maintenance of the roads from Staunton to Greenbrier River, and from Warm Springs to Huntersville. His first duty would be to repair and rebuild the bridges along the roads and to reinforce the embankments which had been weakened by the constant rainfall. To enable Mason to perform this difficult mission, the governor of Virginia empowered him to use "all the appliances for work and labor that have been in use on the State road, and to hire or purchase as you may regard best, wagons, carts, and teams, and all additional labor you think necessary."[45]

Despite Mason's efforts and the official concern with the preservation of these vital arteries, the roads continued to deteriorate over the next few months. Confederate supply lines were slowed by muddy roads on which both horses and wagons became mired. Federals faced the same difficulties in getting supplies to their men. According to one correspondent, "The road from Webster to Philippi, and thence onward to Beverly and Huttonsville, is what is expressively called 'mud pike'." Both civilian and military personnel were employed to provide an army to battle the "oceans of mud and mire."[46]

Meanwhile, both sides continued to probe the other's defenses. About 3 p.m. on September 30 a detachment of about one hundred Union soldiers advanced from Cheat Summit toward Camp Bartow and fired on the outermost stand of Confederate pickets. After returning the fire, the pickets withdrew to the next post. At the bridge over the Greenbrier, the rebel pickets concealed themselves behind a barricade at one end of the bridge. As a detachment of about ten Union troops from the 9th Indiana approached, the Confederates opened fire and a small skirmish erupted. The sound of gunfire reached Camp Bartow and stirred the Confederates into preparations for battle. Colonel Albert Rust of the 3rd Arkansas Infantry galloped out to the picket stand and inquired, "What the hell is the matter here?" Rust then led a small detachment in pursuit of the Federals who quickly retreated. A Confederate soldier from Kentucky wrote, "Whether fright lent wings to flight or not, they can best answer; at any rate their feint didn't pay."[47]

This soldier served in a company from the Bluegrass State under Captain S.V. Reid which was attached to the 3rd Arkansas. These troops wished to return to their native state to guard their homes and families against Union depredations. The Kentuckians grieved when they learned of the hardships endured by secession supporters in their state and pleaded for the opportunity to return home. Rust supported the wishes of the company as did most of his officers. "Yet we are required to remain," wrote the Kentucky soldier. "Why is it? Surely thirty-three will not weaken a command where thousands are in action, when they can accomplish so much for God, justice, and the South in another quarter."[48]

On the first of October, Sergeant John Lyon Hill of the Charlotte Cavalry was ordered to lead another detachment to scout along the Slaven's Cabin road toward the Union fort on Cheat Summit. The troopers were to reconnoiter as far as William Gum's, two miles beyond the blockade at Moore's, even if forced to complete their mission on foot. They arrived at Gum's late in the evening and stayed the night.

At daylight Hill and Private George W. Nicely continued their scouting expedition, finding it "desperate work to make our way over the Slaven's Cabin road." They did not spy any Union soldiers, but they did talk with some local citizens who told them that earlier in September some Confederate militia had routed a Federal detachment sent to eliminate them. During their withdrawal, the Federals had fired upon a small boy and several women, one of whom was Mrs. Gum. A bullet passed through the coat of the boy but he was not harmed. Hill and Nicely returned to camp on the second of October.[49]

As if sensing the impending battle at Camp Bartow, Sergeant James Atkins of the 12th Georgia "walked round the Greenbank road to view the fortifications of logs, stone and earth with cannons sweeping the approach from that quarter for half the range of the guns, the space the road excepted, being completely blocked by trees felled for the purpose." He described the Confederate position as "a strong one, well-suited for slaughtering men." During the day, other Confederate scouts returned from the vicinity of Cheat Mountain Summit and reported that the "Yankees were firing off their guns today," readying for an attack on Camp Bartow. Hill's diary entry undoubtedly reflected the sentiment of many of his comrades: "Let them come. We are ready!"[50]

Preparations for the Union advance intensified. The 15th Indiana, which had marched to the Union fort on September 25, drilled for a few hours on October 1, the first time they had done so in two months. That night the 17th Indiana arrived. Reynolds himself reached camp on the following morning accompanied by fifteen cavalrymen. A short time later both the 7th and 13th Indiana Infantry regiments arrived at Cheat Summit followed by nine pieces of artillery, one company of rifled cannon and one company of brass pieces, all of which were either six or twelve pounders.

The Union troops were exuberant that an attack would finally be made. For the first time in several months they were going on the offensive. Typical of their enthusiasm were the arrangements made by the 13th Indiana, which had begun their march to Cheat Summit on the first day of October. At 7 a.m. the tap of the drum summoned the troops into line, armed and equipped, with one day's provisions in their haversacks. After an inspection, the lieutenant colonel made a brief speech and provided each man with two "Segars." The march then began. As the band joyously played "Yankee Doodle," each man puffed out a cloud of smoke "as though he had been converted into a miniature locomotive."[51]

CHAPTER THREE
THE BATTLE OF GREENBRIER RIVER

On October 2, the Federal camp on Cheat Mountain Summit was alive with activity as the men prepared for the expedition against the Confederates at Camp Bartow. One soldier was unconvinced the march toward Camp Bartow would finally be made, "as we have made every preparation twice and each time the order has been countermanded. If we do go you may expect to hear of some heavy fighting or some tall running on the part of the rebels."

Despite his doubts, spirits were high as the men packed knapsacks, filled their haversacks and cartridge boxes, and zealously cleaned their weapons until the metal glistened from vigorous polishing. The campfires burned bright until about 9 p.m. Around them gathered groups of anxious soldiers, singing, laughing, and discussing the anticipated battle. The glowing fires were extinguished a little before ten, and taps sounded fifteen minutes later. Some of the men tried to sleep, but most were too excited with the expectations of the long day ahead.[52]

A detachment of the Ringgold Cavalry had arrived at the Union fort at nightfall. The troopers filed into an open field where they set up camp and attended to their mounts before preparing a late supper. As they began to eat their standard bland meal, a cow was unexpectedly discovered grazing in the field. To these hungry soldiers the sight of the bovine was "like the ram that was caught in the thorns when Abraham was about to offer his son, Isaac, it must have been there for a purpose; at least we thought it was there for us, and after some difficulty, we brought it down with our revolvers, and soon had it dressed and portions of it roasting on sticks, and thus secured a supper that was very much enjoyed by all."[53]

A short time later a drum roll sounded, bringing the 32nd Ohio into line. They would serve as the advance regiment for the reconnaissance. As the strains of "Hail Columbia" filled the air, the 32nd, led by Colonel Thomas Ford, shuffled slowly off the mountain into the pitch dark night. Accompanying Ford's regiment was the detachment of the Ringgold Cavalry from Pennsylvania and a gun from Captain Philip Daum's Virginia artillery company. This advance force was led by a daring scout from Illinois, A. F. Nicholas. There was then a brief silence atop Cheat Summit.

A little over an hour later, the drums again sounded, this time calling out the 9th and 14th Indiana and the 24th Ohio Infantry regiments. Colonel Nathan Kimball, who commanded the 14th, took time to visit each of his companies, speaking words of encouragement. Kimball, who was born in southern Indiana on November 22, 1822, taught school in Missouri and practiced medicine before the Mexican War, when he served as a captain in the 2nd Indiana. When the Civil War erupted, Kimball became colonel of the 14th Indiana. About thirty minutes after the three regiments filed down the mountain, the 17th Indiana, under the command of Lieutenant Colonel John Wilder, formed in line and was joined by Captain Loomis's Michigan artillery, Howe's battery of United States artillery, Daum's cannon, and a detachment of cavalry.[54]

The mountainside became peaceful once again, but in a short while the sound of a cornet summoned the two regiments that would bring up the rear of the column, the 15th Indiana and the 24th Ohio. They began their march at about 2:30 in the morning, and at last the entire Union column was in motion. The soldiers trudged through the fog while a misty rain lightly brushed their faces. The pace was "snail-like," as the road was narrow and uneven, causing horses to stumble and slowing the forward progress of the artillery. One Federal noted, "the road was overhung by mighty pines, the boughs of which almost interlaced, making the darkness of a very dark night well nigh jet black." At times the troops could continue only by hanging on to the coattail of the soldier preceding him.

The 32nd Ohio had been designated as the advance regiment despite having been in the area only one week. As the troops of the regiment marched along in silence, the stillness of night was broken by the cries of what some believed to be a great horned owl. The sound of the owl followed the column before fading away in the distance. The soldiers soon became convinced that the "owl" was in fact a group of Confederate pickets sending warning signals that the enemy was approaching. For some reason the cavalry detachment accompanying Ford's advance guard refused to head the column. Despite their obstinance, the 32nd Ohio continued until Colonel Ford ran into a treetop near the Gum Road. He was unable to extricate himself and ordered a halt to the march.[55]

Union Colonel Nathan Kimball of the 14th Indiana Infantry. Library of Congress, Prints & Photographs Division, Civil War Photographs [LC-DIG-cwpb-05413]

The 9th Indiana, commanded by Brigadier General Robert H. Milroy, arrived at the Gum Road soon after the 32nd Ohio, and Company H was ordered forward to serve as skirmishers. This advance unit proceeded without opposition until just after daylight, when it reached the first bridge over the Greenbrier. Waiting for them at this point was a detachment of Confederate pickets. As the head of the company approached the bridge, they were greeted with a fusillade. Two soldiers from the 9th were hit: Private Lewis E. Smith was killed, while Private Isaac Bryant was wounded in the shoulder. Before orders could be issued, the head of Milroy's regiment lunged onto the bridge, pouring a volley into the Confederates, killing or wounding three and putting the rest to flight. Troops from both sides discarded blankets, knapsacks, and any other hindrances to the exciting footrace. While some of the Confederate pickets sought refuge in the mountain thickets, the rest were pursued for over a mile before Milroy was able to rein in his troops to await the arrival of the remainder of the column.[56]

Meanwhile, Company G of the 32nd Ohio, led by Captain William Hamilton, was ordered to patrol in advance of the regiment, which was vigilantly guarding the Gum Road. Hamilton moved his company forward about one mile, where the troops halted. The captain then ordered them to take position behind some fallen timber at the base of the mountain overlooking the road. Here they nervously awaited signs of the enemy.

Gunfire was soon heard on the Staunton road and one alert soldier softly murmured, "There they come!" The rest of the company saw that he was right, for three men could be seen making their way forward through the thick underbrush. The order "make ready!" was whispered through the ranks as the company prepared to spring the ambush. Captain Hamilton stepped forward with pistol extended and ordered the three men to halt. The surprised soldiers complied readily after observing that about one hundred guns were aimed at their heads. When asked what regiment they were from, the "prisoners" grinned sheepishly and revealed to the bemused Ohioans that they were members of the 9th Indiana.[57]

The sounds of the skirmish were heard by the other Union regiments marching down the mountain. As the 24th Ohio moved into the lush Greenbrier Valley, the rising sun led the troops to shed their overcoats, which soon dotted the turnpike for miles. The 25th Ohio, which was the last regiment to reach the valley, had not yet received their winter coats, and to these soldiers the appearance of the warm garments was as timely as manna from heaven.

The members of the 25th who scooped up the discarded coats guarded them throughout the battle and carried them back to camp. To the shock and anger of these men, on the day after the battle, brigade commander Colonel Nathan Kimball ordered the coats returned to the 24th, which had apparently complained about the "finders, keepers" incident. The men of the 25th grumbled, but most of the coats were returned. Some, however, were buried in the mountain soil, their one-day owners deciding that if they couldn't wear them, no one would. The regimental history of the 25th recorded the following on the incident: "It seemed to us that the 24th Ohio pleaded the baby act in the overcoat business, and from that time on there never was any real warm friendship between the two regiments."[58]

During the latter part of the march, the road turned muddy and treacherous. The 7th Indiana halted just before daylight and rested on the bank of a deep valley. Private Albert Holmes from Franklin, Indiana, took a nasty fall, tumbling about fifteen feet before landing on a small rock shelf, which saved him from certain death. Holmes lay motionless for a few moments before he could move or speak. A few of his comrades tied some blankets together, lowered them, and slowly hauled up the Union soldier, who was miraculously unhurt.[59]

Camp Bartow, meanwhile, was peaceful on this chilly morning as the soldiers went about their morning routines and duties. The men of Company K of the 31st Virginia were relaxing on their camp stools, preparing breakfast, and awaiting the taste of their morning coffee. Company K had the dubious distinction of having been the first company to retreat at Philippi. Their leader was Captain Henry Sturm, a six-foot, 66 year-old veteran of the War of 1812, who owned a great deal of land in Barbour County and served at various times as justice of the peace, constable, sheriff, assistant United States marshal and delegate in the Virginia General Assembly. Suddenly a breathless recruit rushed up minus his hat and coat, exclaiming to Sturm that the enemy was approaching. Despite the skeptical response the recruit received from the men of Company K, he insisted, shouting, "By jing, they are shootin' our poor fellows at the far river."[60]

At 6 a.m., a messenger from the pickets arrived at the Confederate camp confirming the soldier's report of the imminent approach of a Union

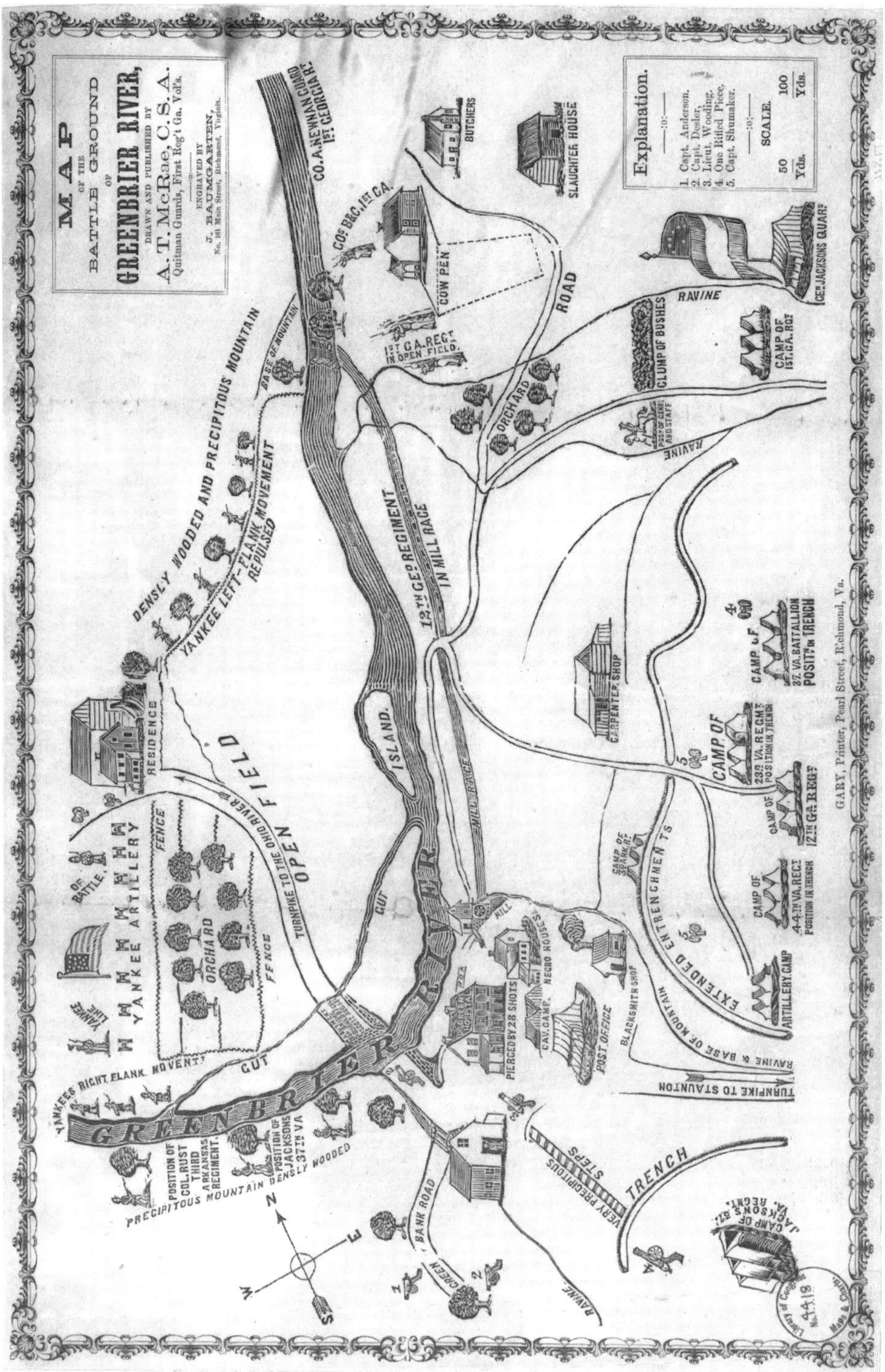

Map of the Battle of Greenbrier River, drawn by A. T. McRae. Library of Congress

column. When Confederate General Henry R. Jackson determined that the enemy was advancing in force, he made plans for the defense of Camp Bartow. Jackson had about 1,800 troops on the day of the battle. The 1st Georgia Infantry, which was to serve as Jackson's reserve, anchored the far right of the Confederate defenses. Next to this regiment was the 12th Georgia.

The center of the line was held by the Fifth Brigade, under the command of Colonel William Booth Taliaferro. Taliaferro, born in Gloucester County, Virginia, on December 28, 1822, graduated from William and Mary and studied law at Harvard. Commissioned a captain in 1847, he was promoted to major of the 9th Infantry before the war with Mexico concluded. Taliaferro also served in the General Assembly and commanded the Virginia militia at the time of John Brown's raid. When civil war erupted, he took command of the 23rd Virginia Infantry.[61]

Taliaferro's brigade consisted of the 44th and 23rd Virginia Infantry regiments and the 25th Virginia Infantry Battalion, better known as Reger's Battalion. The left flank, commanded by Colonel Albert Rust, was made up of Rust's own 3rd Arkansas Infantry, the 31st Virginia Infantry, and Hansbrough's Battalion (9th Virginia). Two pieces of artillery were placed on this flank, in position to enfilade the Huntersville road. Captain Pierce B. Anderson of the Lynchburg Artillery commanded these guns, which protected against an assault from the rear. Four artillery pieces controlled by Captain Lindsey Shumaker were emplaced along the right and center of the defenses as was one gun under the command of Captain William Rice. The few mounted troops stationed at Camp Bartow moved east along the valley to detect and warn of any attempt by the Federals to outflank the Confederate right.[62]

Meanwhile, the retreating Confederate pickets headed by Colonel James Ramsey, field officer of the day, halted a little over a mile west of Camp Bartow. Jackson needed time to get his troops into position so he ordered Colonel Edward Johnson, with a company of the 31st Virginia and another from the 1st Georgia, to assist the pickets in slowing the Union advance. Johnson, a West Point graduate and veteran of the Mexican War, marched rapidly to their relief bringing the holding force to between one and two hundred men.

Taking an advantageous position astride the turnpike, the Confederates awaited the approaching Federals. A few minutes later the head of the Union column marched into view and Johnson's men opened up with a "galling fire." The soldiers of the 9th Indiana, stunned by this deadly reception, quickly recovered and returned the volley. The fighting began in earnest, with the crackling sounds of gunfire reverberating around the base of the mountains. At least one Confederate soldier, Private Jack Munford of the 31st Virginia, was killed in the fighting. Munford was an Englishman who had served in the French Army at Sevastopol during the Crimean War. Hall wrote, "He was shot with a large minie rifle ball, directly through the head—the ball entering his face by the side of his nose. I never saw such a corpse." Johnson urged his troops to hold firm, and hold they did, stalling the Union advance for nearly one hour.[63]

Colonel Edward Johnson was described by one of his men as "about 5 feet 6 in. high, will weigh about 175 or 80: look[s] like [an] old warrier." Another noted after the battle that he was "a brave man, as brave as ever lived." The courage of Johnson, known to his men as "Old Blucher" or "Meat Ax," was prominently displayed during the heated skirmish and his fearlessness was an inspiration to his small band.

The Confederates at Camp Bartow cheered Johnson's detachment even though their view of the skirmish, at first quite clear, quickly became obscured by the smoke. Friends of Private James Hall of the 31st Virginia, who was among the Confederate pickets, told him after the battle's conclusion that "the clouds of smoke rolled up from the valley like a city was on fire." Hall and his comrades also had difficulty seeing as the thick smoke engulfed the valley, but the stubborn resistance displayed by the Confederate skirmishers brought the 9th Indiana to a halt. Johnson took advantage of this brief interlude to send a request to General Henry R. Jackson for artillery support. Captain Shumaker ordered Lieutenant George Washington Wooding to move one piece of artillery across the river to aid the embattled pickets.[64]

While the Confederate detachment temporarily stalled the 9th Indiana to the right of the turnpike, the 14th Indiana, led by Company C, moved into position on the side of the mountain on the left to clear a position for Loomis's battery. Colonel Johnson, seeing the futility of further resistance, finally ordered his men to withdraw. The Confederate advance guard retreated to Camp Bartow in good order, although some were captured by the advancing Federals. These soldiers were forced to witness the battle from the Union perspective.

During the withdrawal Private Solomon Poe of Hansbrough's Battalion was knocked down by the concussion of a Union shell, rendering him nearly unconscious. His retreating comrades thought he had been killed and reported this news to the men of his company. Poe survived, however, and somehow managed to arise and head toward the Confederate lines. The chilly waters of the Greenbrier brought him back to life and he quickened his escape.

One Confederate private did not escape, and his fighting duties were over forever. Private Daniel W. Brown of the Gate City Guards (Company F of the 1st Georgia) heard the order to fall back, but exclaimed, "I will give them one more shot before I leave." Brown, who had already fired twenty cartridges during the skirmish, began ramming another into his gun when he was killed instantly by Union fire.

Colonel James N. Ramsey, who commanded the 1st Georgia, stood near the road and ordered the remaining pickets to "take to the bushes and secrete themselves." Ramsey and several others were cut off from Camp Bartow and did not return until after the fighting concluded, although Ramsey's horse galloped into camp without him during the fighting.[65]

A large Union infantry force advanced and "unfurled to the morning breeze from its centre a large and most beautiful flag with the old stars and stripes emblazoned on it which they 'flaunted insolently in our front'." According to one admiring Confederate, the Federals approached "in a most beautiful line as well 'dressed' as if they had been on evening parade." Several Union batteries took position on both sides of the pike and began to bombard Camp Bartow. A member of Confederate Captain William H. Rice's battery wrote, "We let them fire till they shot some 25 or 30 rounds and then they came up nearer to us and we opened on them."

A Union soldier described the lengthy exchanges that ensued as "awfull and grand at the same time." According to one Union soldier, the noise of the guns in the narrow valley "fairly shook the mountains upon either side almost deafening us with their echo." He added that one "old soldier who was in Walker's filibuster party, and at the storming of Rivas says he never heard heavier cannonading in his life." The balls "came uncomfortably near me. One struck the rail of a fence behind which I lay scattering the rails and the men also and while going up the hill another struck within six inches of my feet covering me with mud." Another ball landed between the feet of Private Joseph Sill of the 25th Ohio, knocking him to the ground and covering him with dirt, but leaving Sill otherwise unscathed.[66]

Sergeant J. Urner Price of the 14th Indiana was not as fortunate. Price was a graduate of Union College in New York and pastor of the Universalist church in Terre Haute, Indiana. He experienced a premonition of his death but calmly advanced into battle. As the artillery duel heightened, Price was struck near the waist by a fragment of shell and was carried from the battlefield. His leg was amputated at the hip and he would be taken to Cheat Summit on the following morning. Here he died, with General Joseph J. Reynolds at his side. According to a notation in the casualty list, Price "died a Christian as he had lived one." Nathan McCann and John Riddle, both in Company B of the 24th Ohio, were mortally wounded during the bombardment; McCann's right leg and Riddle's left were shattered by cannon balls. Both died three hours after being wounded.[67]

At Camp Bartow, General Henry R. Jackson ordered Captain Shumaker to prepare his battery of four guns for battle. Lieutenant Wooding's gun, which had been sent to the aid of Johnson and his men, was positioned in front of Traveler's Repose. The Confederates' one rifled piece was positioned on a hill behind Traveler's Repose, recently vacated by Captain Pierce B. Anderson's gun. Before he could position his remaining artillery, Shumaker was informed that a ball had lodged inside the rifled cannon near the muzzle, rendering it inoperable. Shumaker quickly rode to the scene and ordered the gun withdrawn to the rear where the troops succeeded in ramming the ball into place.[68]

By this time, the tempo of the Union bombardment had increased. Captain Shumaker ordered forward a bronze six-pounder which was placed in the former position of the rifled gun. Soon this cannon began answering the Union artillery, opening up on Loomis's guns, which were posted in an open meadow about eight hundred yards ahead. Captain Howe answered by moving three additional Union artillery pieces forward of Loomis's battery and shelling the center of the Confederate position.

The cannonading focused on a six-pounder manned by a section of the Lynchburg battery. The artillerists "could not stand the fire" and soon fled. Adjutant C. S. Morgan of Hansbrough's Battalion rushed to Lieutenant

David Poe, whose company was supporting the gun and requested men to man the abandoned piece. Although Poe refused to order any of his men to undertake the dangerous mission, he did permit volunteers to aid Morgan. With Morgan serving as gunner, this cannon was soon in operation, and it effectively silenced Captain Daum's lone piece when its third shot struck an axle and broke a wheel, killing a gunner and two horses. Captain Loomis then directed fire against the "mischievous" gun.[69]

To counter the heavy Federal bombardment, Shumaker recalled the rifled cannon. A few balls were fired at the Union guns, but they continued to jam in the muzzle due to the tight fit and the ease with which the gun fouled. Finally, frustrated and disgusted, Shumaker ordered the gun returned to the rear "out of the way." The piece in front of Colonel William Jackson's tent fired ninety-three times and "did terrible execution." The gun was given the name "Colonel Jackson" after the battle.[70]

The first of two Federal attempts to outflank the Confederate fortifications was made after the commencement of the artillery duel. At about 9:30 the head of a sizeable Union column crossed the South Fork of the Greenbrier and attempted to flank the Confederate positions on the left. Although the Confederates on this flank were in position, Colonel Albert Rust was in the process of placing troops between the Greenbrier and the "terminus of abatis to the right of Anderson's battery on the Greenbank road" when the Union advance ascended the hill and neared the Confederate line. Part of the 3rd Arkansas unleashed a deadly volley that put the Federals to flight. In response, two Union guns under Captain Howe directed "canister and shell in large volumes" against the Confederate left, but the shelter provided by the woods kept casualties to a minimum.[71]

With hundreds of cannon balls flying through the air, it was inevitable that casualties were incurred, particularly among the artillerists. On the Confederate side, Captain William H. Rice's gun was particularly targeted as one of the gunners recorded:

> *We shot 65 or 70 rounds with one cannon, and then a ball struck Frank Graves and tore off his left leg and then we moved to another place, and fired 4 times and then Luther Tidler was killed; a cannon ball struck his head and tore the whole top of his head off. Capt. Rice pulled him back and walked about ten steps when a*

> *cannon ball struck his foot and tore it off; and the same ball tore one of our cannon horses legs off.*[72]

At Camp Bartow, a chicken coop was demolished by a cannon ball, killing one of its inhabitants and freeing the rest. More fortunate was Colonel William Taliaferro's mount, a fine bay that was tied next to the trenches, where it fell fast asleep. Once, when a shell burst almost directly overhead, the horse merely raised its head briefly before resuming its nap. A kitten which had been adopted by a company of the 23rd Virginia ran along the entrenchments during the shelling, looking for a playmate. Whenever a ball landed nearby, tearing up the dirt, the kitten would "jump and gambol around it in playful glee."[73]

Perhaps the first pause in the cannonading took place when a Confederate surgeon, unable to locate a yellow flag to mark the hospital building, raised a white one instead. Some of the Union command believed the flag to be a sign of surrender and sent a messenger forward to investigate. Colonel Edward Johnson met the Union man and after listening to his query gruffly replied, "go back and shoot your damned guns." Shoot their guns they did, and soon the building was riddled with cannonballs. Johnson was noted by soldiers on both sides for his bravery in the midst of battle.

Several Union accounts reported a Confederate officer on a white horse was killed by a ball fired by Captain Daum's artillery piece. This was apparently Colonel Edward Johnson, who was not killed, but had a horse shot out from under him and was slightly wounded by a ball that passed through his pants leg.[74]

Despite the deafening roar of the guns, several Union soldiers fell asleep while awaiting the order to advance. Thomas Prickett of the 9th Indiana wrote, "I can say with truth that I had no fear on the field of battle as an evidence of this fact I went to sleep, after we could do nothing more with our rifles, and was awakened by the bursting of one of the enemy's shells—15 ft over my head."

Sergeant Frank Ingersoll of the 13th Indiana noted that half of his regiment fell "fast asleep, that is honest." He also wrote his sister that "if anyone had told me the day before that a man could have gone to sleep with cannon fireing around him I would have told him he lied but <u>honest</u> one half of our Regt did go to sleep in the thickest of the fight."[75]

A more dangerous episode occurred in the trenches occupied by the 31st Virginia. With some of the Confederate guns emplaced about fifteen feet above the regiment, the noise was deafening and it was difficult to pay attention to potential hazards. A shell fell above the trenches and rolled into the midst of the occupants, Company I, threatening their lives. Colonel William L. Jackson quickly ordered one of his men to throw it out, and a brave soldier, Joe Redusky, grasped the shell and pitched it out of the trench only seconds before it exploded, a heroic act that undoubtedly prevented a number of casualties. His fortunate comrades watched the shell roll down the hill until it burst, showering Jackson with dirt and knocking down a private in the company. No one was injured, but the deadly missile did "tear up dirt wonderfully"[76]

Among the members of Company C of the 31st Virginia were three brothers: John, Dan, and Ed Smith. John Smith had been home on sick leave for two weeks, but he had returned to Camp Bartow on this very day, accompanied by his father. Mr. Smith, who was 73 years of age, was paralyzed on one side and required assistance in mounting and dismounting his horse. As the battle unfolded a Confederate officer stopped John Smith and his father in a ravine behind the right flank of the Confederate works. When Dan Smith learned of his father's presence, he and John Henry Cammack made their way to the ravine, where they discovered Mr. Smith arguing vehemently with the officer. Smith wished to join in the fighting but the officer demurred, insisting, "Mr. Smith, you must not go up that bank for you will be hurt." Smith angrily replied, "What do you mean, sir? My blood is as rapid now as it ever was." Smith's sons had no better luck in changing the mind of the elderly gentleman, but the Confederate officer somehow forced him to remain concealed until the battle ended.[77]

At some point during the artillery exchange, Union General Joseph J. Reynolds summoned his officers to a council of war, the results of which made it evident that he no longer considered feasible a full-scale attack on Camp Bartow.

After the war a biographer wrote that General Robert H. Milroy "strenuously urged an immediate attack upon the Reb. works by our whole force by storm" and claimed that a majority of the other officers seconded his request.

An Indiana soldier recorded that Colonel Nathan Kimball asked permission to lead an assault, but Reynolds refused to allow it, "remarking that he merely came out to reconnoitre the rebel position and draw him out." A newspaper correspondent noted that Reynolds rejected the idea of a frontal assault due to the immense loss of life certain to be suffered. According to one Union soldier, another reason for Reynolds's refusal was that he "had no orders to do so."

Reynolds did agree, however, to an attempt to outflank the Confederate right.[78] This flanking movement failed completely from inception to termination. The lead regiment, the 7th Indiana Infantry, was forced to march across open ground until opposite the center of the Confederate camp. Believing that the Federals were about to attempt a frontal assault, two Confederate guns opened on the exposed Union troops causing them to seek cover behind a fence. Using the fence as protection, the 7th Indiana marched on until they reached the right flank of the Confederate lines, where they halted after meeting two enemy regiments, the 1st and 12th Georgia, which had been rushed forward to counter the Union force threatening to turn their position. Five additional Union regiments that had followed in the wake of the 7th Indiana were forced to await the outcome of the lead regiment's encounter with the enemy.[79]

Three companies of the 12th Georgia were ordered about four hundred yards up the river from Colonel Edward Johnson's headquarters in order to prevent the Union force from outflanking their position. When the officers of the regiment spied the Federals moving through the bushes toward the Confederate right flank, they immediately ordered the remaining companies of the regiment into action.

Led by Johnson, who rode calmly into a deadly rain of enemy fire, the 12th Georgia advanced toward the enemy, according to one observer, "moving in double quick time and in a very handsome style, cheering as they went." Crossing the Greenbrier, which was waist deep, the Georgians continued for about one hundred yards when, upon confronting the 7th Indiana, they were suddenly ordered to withdraw. They re-crossed the river and took prone positions in a mill race, which served as a trench. Their officers ordered them to hold their fire until they could see "the white of their [the enemy's] eyes."

General Jackson rushed to the 1st Georgia, which held the far right of the Confederate defenses, and ordered them into the valley below their

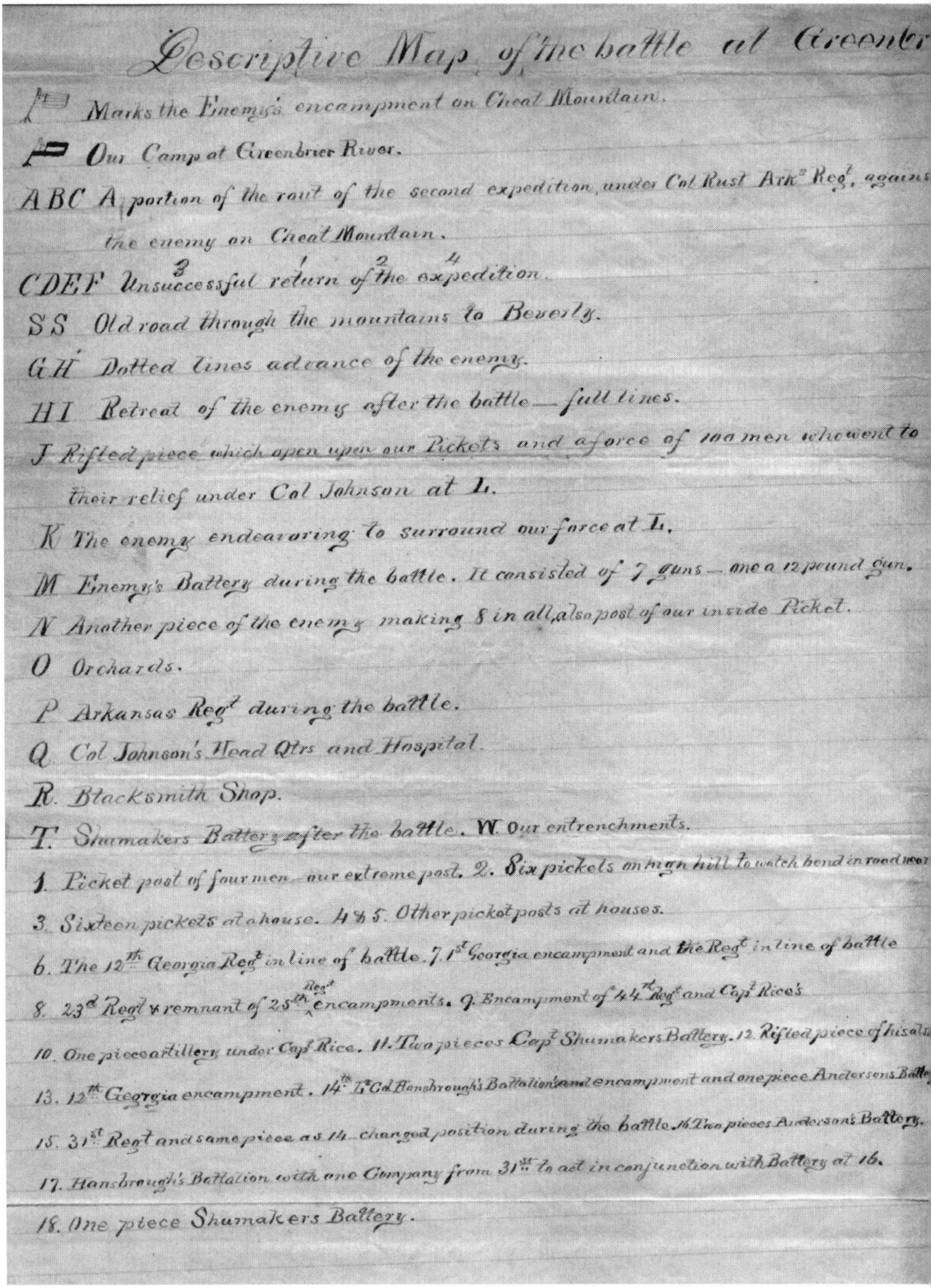

Descriptive Map of the battle at Greenbr

Marks the Enemy's encampment on Cheat Mountain.

Our Camp at Greenbrier River.

ABC A portion of the rout of the second expedition, under Col Rust Arkd Regt, agains

the enemy on Cheat Mountain.

CDEF Unsuccessful return of the expedition.

SS Old road through the mountains to Beverly.

GH Dotted lines advance of the enemy.

HI Retreat of the enemy after the battle—full lines.

J Rifled piece which open upon our Pickets and a force of 100 men who went to

their relief under Col Johnson at L.

K The enemy endeavoring to surround our force at L.

M Enemy's Battery during the battle. It consisted of 7 guns—one a 12 pound gun.

N Another piece of the enemy making 8 in all, also post of our inside Picket.

O Orchards.

P Arkansas Regt during the battle.

Q Col Johnson's Head Qtrs and Hospital.

R. Blacksmith Shop.

T. Shumakers Battery after the battle. W. Our entrenchments.

1. Picket post of four men—our extreme post. 2. Six pickets on high hill to watch bend in road near

3. Sixteen pickets at a house. 4 & 5. Other picket posts at houses.

6. The 12th Georgia Regt in line of battle. 7. 1st Georgia encampment and the Regt in line of battle

8. 23d Regt & remnant of 25th Regt encampments. 9. Encampment of 44th Regt and Capt Rice's

10. One piece artillery under Capt Rice. 11. Two pieces Capt Shumakers Battery. 12. Rifled piece of his als

13. 12th Georgia encampment. 14th Lt Col Hansbrough's Battalion and encampment and one piece Andersons Batte

15. 31st Regt and same piece as 14 changed position during the battle. 16. Two pieces Anderson's Battery.

17. Hansbrough's Battalion with one Company from 31st to act in conjunction with Battery at 16.

18. One piece Shumakers Battery.

Descriptive map of the Battle of Greenbrier River, drawn by C. S. Morgan. Charles S. Morgan Papers, West Virginia State Archives.

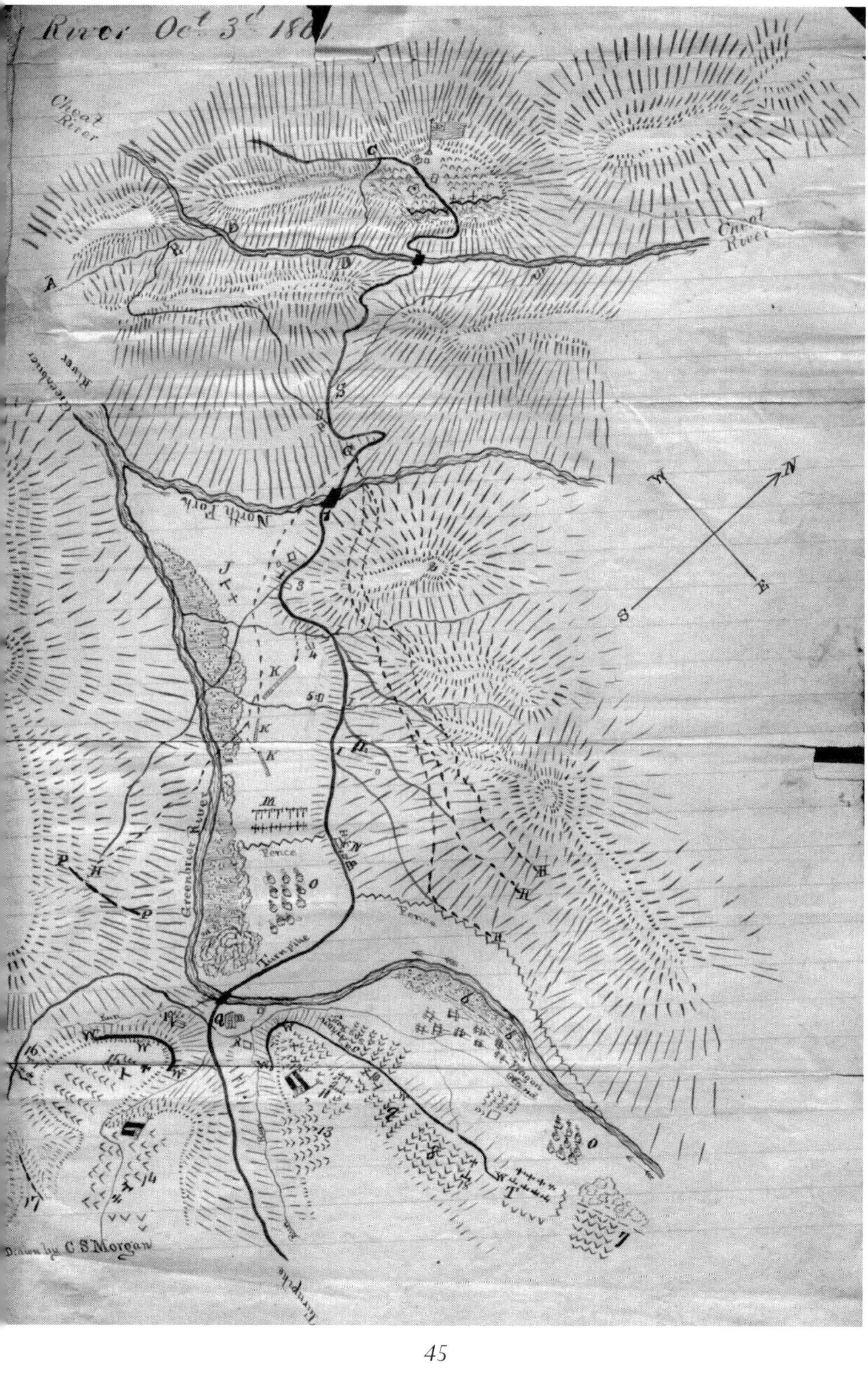
River Oct. 3d 1861
Cheat River
Cheat River
Greenbrier River
North Fork
Greenbrier River
Fence
Fence
Turnpike
Turnpike
Drawn by C S Morgan

camp. He expected a flank attack on the Confederate right and ordered the Georgians to hold this position at all costs. Reaching the field below, they formed in line and then lay flat on the ground to await the Union attack. About 10:30, Jackson dispatched twenty men around the mountain on the Confederate right to determine if other Union troops were attempting to turn their position.

The troops of the 7th Indiana advanced to within four hundred yards of the 12th Georgia before they halted and opened fire "with their long ranged guns," wounding two of the Georgians. Despite the exhortations of their officers, the men of the 7th Indiana refused to move forward and were exposed to a heavy cannonade emanating from two Confederate artillery pieces commanded by Captain Shumaker. Seeing the futility of the flanking movement, General Reynolds ordered the Union regiments withdrawn. Although the intensity of the Confederate artillery fire played a role in repressing the attempted maneuver, the positions of the Georgia regiments were also crucial. As one Georgian wrote, "The posision wee took whiped them without us fireing a gun on them."[80]

The 32nd Ohio had been left behind to guard the rear of the Federal force. As the sounds of battle slowed, the regiment observed the Union column withdrawing from the field. The men of the 32nd were told that the enemy was in pursuit and preparations were made to receive them "in grand style." Captain Jackson Lucy of Company A "mounted a stump about 4 feet high and taking off his hat, held it up above his head, and told his men to be cool and wait until they could see the whites of the rebels eyes, take deliberate aim, and when he dropped the hat, to let them have it." Lucy was praised for his bravery but no Confederate troops appeared, and the column wound its way back up the mountain.[81]

On Allegheny Summit the troops of the 52nd Virginia were ordered under arms at the first sound of cannon fire. From their lofty position nine miles from the fighting the men of the 52nd could hear "with great distinctness every report of the cannon and the rattle of musketry."

Extra ammunition was issued and the soldiers prepared to march down the pike toward the battlefield. Soon a messenger arrived from Camp Bartow, ordering the regiment to guard the road leading to the rear of the Confederate camp. Colonel John Baldwin dispatched a scout who rode six miles down the road, while Baldwin and Major John D. Ross led a company forward to prepare an ambush.

Another messenger arrived, ordering the regiment to be ready to move forward at a moment's notice. The scout soon returned and reported that there was no enemy movement on the Green Bank road. A short while later, yet another dispatch was delivered, requesting the immediate advance of the 52nd to Camp Bartow. Within minutes, Baldwin's regiment began to march down the turnpike, "cheering and singing with drums beating." After descending two miles, however, an aide to General Jackson rode up and informed Baldwin that the battle had ended in a Confederate victory. The troops of the 52nd were terribly disappointed at not being able to participate in the fighting. Major Ross wrote, "You never saw as crestfallen a set of chaps in your life; a great many of them expressed their willingness to give $5.00 to be there, which considering the low state of their finances is a large sum." Jackson's aide ordered the regiment to remain in place and to advance if they heard further gunfire. To their chagrin, the battle was indeed over, so they returned brokenhearted to their camp on Allegheny Summit just before sundown.[82]

Possibly their march down the mountain was observed by Reynolds and his men, since he and several other Union soldiers recorded shortly after the affair that the Confederates were being reinforced near the conclusion of the battle. As is common in any war, estimates of enemy strength were wildly exaggerated. Sergeant Frank Ingersoll of the 9th Indiana believed the reinforcements increased enemy strength to 15,000. The pending arrival of the 52nd Virginia may have been the final straw for Reynolds, although the fact that ammunition for the artillery was low was also a significant factor in the decision to withdraw.[83]

Late in the afternoon many of the Confederates explored the positions held by the Union troops during the battle. One soldier recorded that the earth was "literally plowed up with our cannon shot." Cannon balls had pierced "that house on the river bank" near the encampment of the Churchville Cavalry and torn through the tents of the troopers. Discarded canteens, blankets, and bayonets were among the less gruesome finds. "Much blood was seen. Several arms and bones were found along the edge of the field where they had drawn up their infantry," observed one Confederate.

Also discovered were the bodies of several Union soldiers, left by their comrades on the battlefield. Private James Hall of the 31st Virginia was particularly fascinated by one Union fatality, Private Albert I. Abbott of the 9th Indiana, whose death was attributed to the force of a cannon ball. According to Ambrose Bierce of the 9th, Abbott "was lying flat upon his stomach and was killed by being struck in the side by a nearly spent cannon-shot that came rolling in among us. The shot remained in him until removed." Bierce was informed that "it was a solid-round, shot, evidently cast in some private foundry, whose proprietor, setting the laws of thrift above those of ballistics, had put his 'imprint' upon it; it bore, in slightly sunken letters, the name 'Abbott'."

Discovered in the Union soldier's portmanteau were three twists of hair, treasured gifts from his sisters. Also found were furlough papers and a letter to his sisters, which would never reach Indiana. In the letter Abbott explained that his return home would be delayed long enough for him to descend the mountain and participate in an anticipated victory, so that he would have tales of glory to share with them upon his return to the Hoosier state. Instead, as Hall crudely noted, Abbott's furlough had turned into a "fur-long."[84]

Casualties in the battle were surprisingly light considering the numbers of troops engaged, yet not so unexpected since most of the fighting was done by the artillerymen. Fifteen Union soldiers were killed or died shortly after the battle from wounds received in action; thirty-five others were wounded. Nine Confederates were killed or died from wounds while thirty-four were wounded. Sixteen others captured during the battle were described as "hard looking chaps."

Sergeant David Beem of the 14th Indiana, along with Jess Herrold and other members of his company, captured seven Confederates. Beem seized "a splendid rifle gun, and a beautiful pistol, which I took from a rebel officer. The gun and pistol are worth 40 dollars." Reynolds's men also captured fifteen horses and eighteen head of cattle.

Among the Confederate prisoners was William Slayton, whose brother John had served as a spy and guide for the Union troops at Cheat Mountain since their arrival in the region. As a Union captain noted, "Thus you see, brother has been arrayed against brother." While the small number of Confederate casualties was due in large part to their entrenched positions, the modest casualties incurred on the Federals was attributed to the inexperience of the Confederate artillerymen. Colonel John T. Wilder of the 17th Indiana wrote that "not more than one of their shells in ten burst, they forgetting to cut the fuses," while Captain Elijah H. C. Cavins of the 14th Indiana similarly noted, "their new hands didn't know how to fix their bombs, and the consequence was that they didn't explode."

The Union artillery was somewhat more effectual, as in the Confederate camp "nearly every tent had a hole in it and one burnt clear up." Another Confederate noted that "tents are riddled with balls and shell."[85]

The 7th Indiana was vilified for alleged cowardice for their actions along the Confederate right flank. These accusations stemmed in large part from the humiliating experience of having lost their regimental flag during the battle, not by the force of arms, but rather the sweet arms of slumber. When the 7th had advanced to support the artillery at the outset of the fighting, the men were forced to hug the ground to avoid the heavy shelling. The color bearer of the regiment leaned the flag against a fence corner and soon fell asleep. When he was roused to join the movement on the Confederate right flank, he forgot to grab the colors, which were left on the field, later to be "captured" by the Confederates.

This incident led to a nickname for the 7th: the "Banner Regiment." In response to the accusations, General Robert H. Milroy wrote a letter to the *Cincinnati Daily Times*, stating that the "7th Regiment upon that occasion acquitted themselves honorably and were guilty of no disgraceful conduct whatever. The charge is entirely groundless." The letter was also signed by Colonel Jacob Ammen of the 24th Ohio and Lieutenant Colonel William P. Richardson of the 25th Ohio. The captured flag, "discovered in a laurel thicket," was paraded through the Confederate camp, producing "many a lusty cheer." Described as "a new flag, about eight by ten feet in size," it was delivered to the War Department in Richmond by Lieutenant W. D. Humphries, aide to General H. R. Jackson.[86]

Despite the large number of troops present at the Battle of Greenbrier River, the engagement of October 3, 1861, was almost exclusively an artillery duel which did nothing to alter the fall stalemate in the mountains of western Virginia and was apparently not part of a concerted Union plan to advance eastward along the Staunton-Parkersburg Turnpike.

The Confederates claimed a great victory, but they believed the Federals would make another attempt to dislodge them and renewed their efforts to fortify the camp, which controlled the vital turnpike leading over the Alleghenies. Although Union General Joseph J. Reynolds and his men believed they would launch another attack, the Federals of the Cheat Mountain District returned to their various camps; they would not again in 1861 have a concentrated force as large as the one which participated in the engagement.[87]

The Union troops were disappointed by the failure to dislodge the enemy at Camp Bartow. Private Lucian Barbour of the 14th Indiana summed up his part in the expedition in a direct and humorous way: "I marched 30 miles through mud & rain, went double quick four miles, fought six hours, was knocked down with a cannon ball, & all this without breakfast or dinner—enough to kill a mule."

Others were frustrated by Reynolds's refusal to order an all-out assault on Camp Bartow. The general's official report stated that the main objective of the expedition had been accomplished, but others disagreed. Private Alf Welton of the 9th Indiana recorded, "what should have proved a victory turned out a mere farce." Welton also believed General Milroy and "all other fighting officers were very much disgusted at the manner in which the battle! was (not) conducted." Another soldier wrote that he and his comrades were "all certain could easily have taken possession of the camp and much disappointment is expressed at the failure. I expect the General knows why he didn't, and as he knows more about war than I do, I shall acquiesce, hoping that he was, and that time may prove it."[88]

Reynolds was undoubtedly correct in refusing to order a direct assault. The Confederate works were impressive and a frontal attack would have resulted in numerous casualties. The ground in front of the camp was level and open, meaning that the Union attackers would have been easy targets. The planned attack on the Confederate right flank might have been successful, but the Union soldiers and their officers were apparently uncertain as to their objective and the assault never developed. Had Reynolds been able to maneuver a few pieces of his own artillery to this position to cover an advance, it seems possible the Federals could have successfully turned the Confederate right.

One unanswered question about the battle concerns the objective of the Union advance on Camp Bartow. After it was over, Reynolds, his officers, and troops referred to the engagement as an armed reconnaissance; yet, there is evidence that the advance was a planned attack that failed to develop due to the unwillingness of General Reynolds to risk an assault on the well-entrenched Confederates.

General Robert H. Milroy recorded after the war that, on about September 27, Reynolds "consented to go over and attack the Reb forces." Not enough evidence exists, however, to determine conclusively that the original intent was to launch a full-scale attack. One Union soldier summed up the confusion: "The attack was made more with a view to ascertain the enemy's strength than to take his position. It was an armed reconnaissance except that we attacked them."[89]

In retrospect it appears that if Reynolds had planned to fight rather than reconnoiter, he could have flanked the Confederate position by marching down the Gum Road to Green Bank, then north to attack Camp Bartow from the rear. This route was undoubtedly longer, and the roads were more primitive than the turnpike, but if a couple of regiments had feigned an attack from the pike while the main body swung in behind the Confederates, success would have been more probable.

In all likelihood, however, Reynolds had no intention of capturing the enemy camp on the Greenbrier River. Instead, he was simply following the directive from Rosecrans: "If your troops move lightly, carrying the necessary provisions to go and return, you can so harass them that you can prevent their leaving their position and stop any force they can bring against you."[90]

The Confederate victory stemmed from two factors. First, the strong position and defensive works of the Confederates not only shielded the troops from much of the shelling and afforded Confederate gunners a clear field of fire, but also intimidated Union troops and officers. Although the construction of the works at Camp Bartow was not entirely completed, they were nonetheless quite formidable at the time of the battle. Trenches lined the hillsides on both sides of the roads and abatis had been placed around the junction of the Green Bank and Staunton roads. Significantly, trenches and gun emplacements can still be seen at the site today, an indication of the engineering skills of Lieutenant Colonel Seth Barton and the labor exerted by the troops who constructed them.

The second reason for the Confederate victory was the restraint exercised by Union General Joseph J. Reynolds. Although Reynolds had a notable edge in troops and artillery, he was simply unwilling to attack the well-entrenched Confederates. Both attempts to outflank the Confederate camp were utter failures, and Reynolds had no hope for an easy victory and no will to attempt to gain a costly one.

It is probable that the Union command in western Virginia was content with holding its early gains and unwilling to risk a major engagement that might reverse Federal control of the region. Reynolds's superior, General William S. Rosecrans, withdrew from a confrontation with Confederate General Robert E. Lee at Sewell Mountain without risking an attack and apparently Reynolds displayed similar caution, although it is unclear whether this caution was concerted.

Perhaps the best explanation for the failure of the Union advance on Camp Bartow was provided by a regimental historian who wrote,

> *What the main object of the expedition really was, will now never be known. We had enough material, eager for a fight, to have easily taken the position, and in either regiment there were half a dozen subaltern officers, who with the next two years' experience, would unhesitatingly have attacked and carried such works with half the number of troops we had on the field. It was early in the war, and the fruit had not yet matured.*[91]

Confederate General Henry R. Jackson issued a congratulatory address to his men two days after their successful defense of Camp Bartow:

> *Soldiers: After a campaign, to you of peculiar hardship, the enemy, descending from his mountain fastnesses, has afforded you the long coveted opportunity of testing your efficiency in action. Confident in his greater number, his superior arms, and the comparative weakness of your position, he came, with his wagons trains, in anticipation of an easy victory and a permanent enjoyment of its profits. But gallantly and well have you maintained your place in line with your brothers of the army of Northwestern Virginia. Meeting his earlier advances with striking exhibitions of individual daring, receiving his concentrated fire for more than four hours with the coolness of veterans, and then, when he supposed your spirit to be shaken, with a calm determination, indicating to him what his fate would be should he attempt to carry out further his*

original designs, have you repulsed his first efforts to charge and to break your lines. Though you were not in force to pursue him, and to realize the full fruits of your triumph, retreating he left behind him the unmistakable evidences of his rout in the bodies of his dead, too numerous to be carried with him, the abandonment of arms and accoutrements, even to the colors which in the morning he had flaunted so insolently in your front.

Soldiers! With pride I congratulate you.[92]

CHAPTER FOUR
A LIVELY MONTH

Many of the Confederate soldiers at Camp Bartow spent October 5 relaxing, writing letters, and generally basking in the afterglow of victory. Due to his heroic leadership, Colonel Edward Johnson's reputation was transformed, and he had "become the most popular officer here." Some of the men regaled their comrades with tales of their own heroism during the battle.

For others it was business as usual. Sergeant John Lyon Hill accompanied a detachment of pickets who advanced to Slaven's Cabin. They found that the old house had been converted into a makeshift hospital by the Federals after the battle. Hill discovered "pools of blood, bloody clothing and men's fingers which had been shot off."

On the same day, a tragedy occurred at Camp Elkwater when Lieutenant Solomon J. Bidwell of the 6th Ohio Infantry was accidentally shot and killed by a member of the 17th Indiana. This soldier was crossing a stream when a water snake darted between his legs. Impulsively, he raised his weapon to fire. The lieutenant, who was standing on the bank, exclaimed "Don't shoot!," but the Indianan ignored him, and the bullet ricocheted and struck Bidwell in the breast. He died within minutes.[93]

Work on improving the fortifications at Camp Bartow continued on the following day, during which a small Federal party bearing a flag of truce advanced to the outermost Confederate picket post. A note was sent to General Henry R. Jackson requesting the bodies of the Union troops which had been killed and left on the battlefield during the hasty retreat but Jackson refused to accept the note because it was improperly addressed to "the officer commanding."

Colonel Edward Johnson met the party and told them they could not enter the camp. He also informed them that "he had buried their dead and that he could bury the rest of them if they would come down." One of the Union soldiers said that "he had a brother that was col. of a regiment in our army some where in Virginia. That is an awful thing for brother to be fighting against brother." Jackson later relented, agreeing to bury the Union troops in coffins. He refused, however, to allow any Union messenger into the Confederate camp.[94]

Meanwhile, the Union troops had returned to their respective encampments following the failed attack on Camp Bartow. The Cheat Mountain Division of the Army of Occupation, Western Virginia, led by General Joseph J. Reynolds, was formed into brigades on October 7. General Robert H. Milroy would command the 9th Indiana, and the 24th, 25th and 32nd Ohio Infantry regiments at Cheat Mountain. A second force, comprised of the 2nd (West) Virginia, the 3rd and 6th Ohio, and 7th Indiana, was under the command of Brigadier General Ebenezer Dumont at Elkwater. Reynolds himself headed the third brigade, stationed around Huttonsville. This force consisted of the 13th, 14th, 15th and 17th Indiana Infantry regiments. All three brigade commanders were from Indiana, a fact that rankled several Ohio officers. General Reynolds requested that General Robert Schenk be sent to the region "to quiet them."

Reynolds also notified General William S. Rosecrans, "Enemy at Greenbrier, could probably be taken by my command by getting down in front of them and tarrying a few days but just at this time other pressing necessities crowd upon us." Those necessities included additional transportation, supplies, and rations for both man and beast. Since early August, Reynolds had been "begging" for wagons to keep supplies moving forward across the nearly impassable roads.

Crews worked daily on the roads leading from Webster and Beverly to the camps in the Cheat Mountain District but the continual rainfall made these vital arteries, which were unused to such heavy traffic, muddy bogs. Reynolds complained that his men had "not two days provision ahead and not certain when will be," and that "animals [were] on half rations and many men without bread." An attack on Camp Bartow would not be made "unless I can do it right."[95]

The 13th Indiana was stationed two miles from Huttonsville on the turnpike near Elk Fork and Falls Creek. One soldier described the site as "the greatest place in the state for Trout fishing." At Camp Elkwater gambling had become rampant and led to the forwarding of pay back to the men's home states. One disapproving correspondent observed:

> *I have seen some of the worthies, for want of better convenience in the way of gambling, get the head of a flour barrel for a table,*

with numbers from one to six marked on it with a piece of charcoal, a mustard box for dice-box, seated in some retired spot under the river bank, or in a fence corner, a little outside the camp, working away hard at it, making the old tin box rattle again, a group of ten or twenty looking on or participating in the game, all deeply engaged, the spoonies and green ones getting regularly skinned, as a matter of course.

This correspondent believed that gambling would cease if the soldiers' pay went to their families. "Give him all the regulations call for, of good quality as well as quantity, and he can dispense with the use of money," the writer observed, concluding, "He needs none."[96]

Two days after the battle, Colonel Ike Marrow led the 3rd and 6th Ohio, part of the 2nd (West) Virginia, a company of cavalry, and three pieces of artillery on an expedition to reconnoiter the Confederate camp at Big Spring. The column departed Camp Elkwater at 3 p.m. and began the rugged march through another drenching rain.

On the following day, Marrow's force reached Mingo Flats, the scene of an encounter nine days earlier. While the soldiers partook of a quick meal, Marrow and some of the cavalry rode four miles to Big Spring, the former site of a Confederate headquarters. Here they found tents slashed and army stores destroyed and scattered about the ground. Marrow also found charred barrels that held guns, knives and other weapons burned as the Confederates retreated. A local farmer informed Marrow that they had left shortly after Sullivan's September 26-27 advance on their camp. The impassability of the road made a further advance impossible, so the Union force returned to Elkwater, arriving on October 8.[97]

That evening near Camp Bartow, Confederate pickets heard disturbing cries nearby, and searches were conducted that night. On the following day, Sergeant James Atkins of the 12th Georgia, astride the mount of Colonel Albert Rust, led 22 men of the 3rd Arkansas on a search of the area. Despite a thorough exploration, they were unsuccessful in locating the source of the cries. Later that day, four additional pieces of artillery arrived at Camp Bartow. Two were twelve-pound howitzers; the others were six-pound rifled pieces.[98]

Prices for the luxuries of life, supplied by local citizens, were skyrocketing. Butter sold for forty cents a pound, honey in the comb cost fifty cents a pint, and whiskey ("the meanest kind of that"), perhaps the most important rarity, cost three dollars per canteen.

About nine o'clock on the night of October 9, three Confederate scouts left camp to reconnoiter the Slaven's Cabin road. They spent the entire night at Heavener's. Moving out at dawn they went to Green's on Cheat Mountain where they learned that, on the day before, a Union soldier had been spotted in the brush near Moore's. A Mr. Collins provided the scouts with a hearty breakfast of stem peaches, rye bread, and rye coffee. After devouring the timely meal they returned to camp.[99]

October 10 was a "fine" day and most of the troops at Camp Bartow were in high spirits. The men found more than a bit of humor in a New York newspaper report that they had been "completely routed" by the Union advance on October 3. Despite the good cheer, the Confederates expected to be attacked at any moment.

A rumor had reached camp that Union General William S. Rosecrans was retreating down the Kanawha Valley. If true, this would leave the Union force on Cheat Summit in a precarious position. If their supply route was cut off, they would be forced either to retreat or to attack in order to gain a new source of supplies farther east. General Jackson issued Special Order Number 132, directing all Confederates who were not sick or detached on special orders to immediately return to their regiments.[100]

October 12 was cold and wet. A heavy rain had fallen throughout the night but the clouds rolled away by mid-morning. The Confederate soldiers were anxious for the mail train, which had not been seen for two days due to the horrid condition of the roads. Sergeant John Lyon Hill again set out as messenger of the advanced picket post at Arbogast's at the foot of Cheat Mountain. Here he learned that, on the previous night, a member of Captain Rice's artillery company had almost been shot by the pickets. This soldier had been in hiding near Beverly since the Confederate retreat from Laurel Hill nearly three months earlier. Making his way to Camp Bartow, he had very nearly become a victim of friendly fire when he was mistaken for a Union soldier. Fortunately, the pickets held their fire long enough for the soldier to identify himself. Hill spent the night of October 13 at Arbogast's and complained that he "slept but little in consequence of the cold weather and the proximity of the Yankees."[101]

General Jackson fired off a letter to Richmond on October 12 complaining about the inadequate supply system, which was becoming increasingly worse as the deadly winter approached. Jackson was particularly critical of Major James Corley, the quartermaster for the Army of the Northwest, who had not made an appearance for over two months. Corley's lack of attention resulted in numerous shortages, including an acute scarcity of forage for the horses, rendering many of them unfit for service.

Jackson wrote, "The cold weather is already upon us, and with a winter climate like this, with more than 70 miles between the army and its depot of supplies, with the connecting road becoming daily worse and worse, and with the country immediately adjacent already exhausted, unless new energy be infused into the staff department it is not easy to conceive the amount of loss, embarrassment, and suffering which must in a short time be developed."[102]

Jackson had about thirty Charlotte and Rockbridge cavalry stationed in the South Branch Valley to keep an eye on Union activity in the region. Lieutenant F. G. McCoy had been sent to Petersburg with a small force, but on the day after their arrival they were attacked and driven from the town, losing nearly all their baggage.

McCoy and his men encamped at Tract Bridge, about fourteen miles from Petersburg. Soon after the move, many of the Confederates fell ill with typhoid fever, measles, or jaundice. Within a short time six of McCoy's men had died of disease and another had been wounded. Although his command remained understrength, they had still managed to capture fifteen or twenty Union men, who were sent on as prisoners of war.

In mid-October, McCoy ordered out a party of thirty mounted men and twenty volunteer militia to the Dry Fork/Seneca region to look for Union guerilla John Snyder. This detachment wounded one Union soldier and killed another during the expedition. On two occasions after this scout, McCoy marched into this region with his entire available force, scouting the "Union Country" in the northeastern portion of Pendleton County and the southeastern part of Hardy. His scouts traveled as far as Petersburg and Moorefield but could not venture any farther due to the danger of being cut off.[103]

Most of the fighting in this region took place between armed citizens, who conducted ruthless warfare reminiscent of the Revolutionary struggle.

Secessionists from Moorefield raided a "Tory rendezvous," capturing some horses and firing into a house where the Union men were meeting. Lieutenant McCoy, who commanded the small Confederate cavalry force, wrote that the raid would only incite retaliation and that "old fashioned Tory warfare" was likely to become prevalent in Hardy County: "I have tried to discourage those predatory attempts on the part of our citizens, because they are not creditable, and because, as affairs now stand here, they are likely to result in injury to the true people generally of the county; and to the cause in other respects."

Although McCoy sought to prevent guerilla warfare, he also aggressively pursued notorious Union sympathizers like John Snyder. On October 14, he ordered a detachment of twenty men, half infantry and half cavalry, into the Allegheny Mountains in search of Snyder, whom he described as "a very noted, active and dangerous traitor." Snyder had been seriously wounded earlier in the year by some of his neighbors, but he was in the process of recovery and had begun venturing out once again.

When the Confederate detachment reached the Dry Fork of the Cheat River they learned that Snyder had been removed from the area by Union troops, but the Confederates reportedly seized six Union sympathizers. On their return they ran into a Union force they estimated at one regiment, which led them to flee precipitously to their camp on the South Branch. McCoy believed that the increasingly frequent presence of Union troops in the Dry Fork region was a consequence of a skirmish at Laurel Fork in August as well as of the attack on Snyder and served to punish the secessionist citizens of the region.[104]

Despite the strength of the Confederate positions, improvements proceeded at Camp Bartow. On October 15, Georgian Shep Pryor wrote, "Wee are now tolerable well fortified: got 12 pieces of cannon and places all fixed for the [men] to shoot from; that is, fortifications for the cannon with openings to shoot through so the men can man the cannon and not be exposed to the enemy while doing it. Wee also have ditches for our infantry to get in so the enemy cant use their long ranged guns to any advantage over our short ranged guns, for they cant hurt us untill they get as close as wee want them. Then wee have an equal showing in guns & the advantage of the ditch to fall and load in. This is now a verry strong place & will in 5 or 6 days be a great deal stronger..."[105]

Early on the morning of October 15, Colonel Samuel Fulkerson's 37th Virginia Infantry Regiment arrived at Camp Bartow. This regiment had been guarding a mountain pass seven miles southwest of the Confederate camp. On this same morning, nine Union companies descended from Cheat Mountain to probe the Confederate defenses. About ten o'clock they encountered the advanced outpost at the foot of the mountain. Upon reaching the bridge, a squad was posted on a boulder, partly concealed by a cluster of bushes. Confederate pickets fired, and the Union troops responded with a volley.

An Ohio soldier wrote, "As soon as the pickets fired, we started on double quick down the hill, we got down soon enough to see them running on double for life." Several cavalry pursued and captured one man who had fired two shots at a captain. The remainder of the Union detachment "rushed on through fields and climbing fences till we got in sight of Greenbrier Bridge, where we formed into line of battle and awaited for our scouts to come in. We then went back to the second bridge where our firing first commenced and staid until all came up."

John Lyon Hill, who was scouting in the area, had stopped at Slaven's and was startled by the firing. Hill and the other Confederates immediately formed an ambuscade and awaited the approach of the Union troops, but they had returned to their mountain lair atop Cheat Summit. A note to Colonel Albert Rust was later found posted on the bridge, inviting him to pay another visit to the Union camp at Cheat Summit. The Union detachment returned to camp with their prisoner, "a Hungarian by birth," in tow. Discovered in the captured man's pocket was a copy of the *Richmond Enquirer*, which contained an exaggerated account of the Battle of Greenbrier River. The man was reportedly armed, but "all attempts to elicit information from the gentleman signally failed."

At Camp Bartow, the firing was heard and the soldiers were ordered to prepare for battle. The sick were quickly gathered and moved west up the pike. The 12th Georgia marched up the hill above the rifled gun battery, and they remained in this position until three in the afternoon.[106]

At midnight on October 16, Lieutenant Samuel Dawson of the 12th Georgia, known as a "celebrated Indian hunter," led a Confederate detachment of about one hundred troops to scout along the road toward "Yankeedom." About three miles from Cheat Summit Fort, Dawson's men encountered a large Union force, estimated at one thousand men. After a brief exchange, the outnumbered Confederates retreated. When Dawson's detachment returned to camp in the afternoon, they were fired on by their own pickets. Fortunately, none of the troops were hit, although they were certainly angered by this inhospitable homecoming.[107]

The firing of the pickets roused the troops stationed at Camp Bartow. They were again called to arms by order of General Jackson after a scout arrived and reported that Federal troops had fired on the pickets and were advancing toward the camp. "In a moment," one soldier observed, "the camp presented the appearance of an immense beehive—companies of soldiers were marching in every direction, getting their positions, and all anxious for the fray, for which they have been praying since the battle of the 3d." The camp returned to its previous calm when it was learned that the firing had been directed toward Dawson's detachment.[108]

Despite a rumor that Union Generals Rosecrans and Reynolds had united their forces, the Confederates were confident due to the newly arrived reinforcements, the additional artillery, and improved defenses. "If they do come," one overzealous Confederate wrote, "you may rest assured that the bloodiest battle will ensue that ever took place on the American continent, and there will be more dead Yankees left in the meadow at Traveller's Repose, than the sons of Dixie will be able to bury, or the ravenous vultures of the Alleghanies to devour, in three weeks."

At least one Confederate was more concerned about news and events from home, warning speculators in his home state of Georgia that they would "rue the day that ever they went into such speculations." He also stated that when he returned home, "if I dont get the necessaries of life at reasonable prices my name aint Clayton Wilson."[109]

On October 16, Colonel John Baldwin, commander of the 52nd Virginia, sent a letter to Confederate Secretary of War J. P. Benjamin. Baldwin, who was stationed with his regiment on Allegheny Summit, complained of the lack of provisions and supplies in the area. These shortages forced the Army of the Northwest to draw provisions from the South Branch Valley or from the Staunton area. Shipments from the South Branch region had so often been intercepted by Union troops or loyalists that it was becoming difficult to find teamsters willing to make the risky journey. The road to Staunton was useable during the summer months, but in the winter it would become nearly impassable. Already Baldwin had been forced to detach large crews to maintain it. The horses had been reduced to half rations of corn, hay being almost impossible to obtain.

Baldwin also warned of the lack of preparation for the fast approaching winter season. Although it appeared there would be no withdrawal east of the Alleghenies, no huts or houses had been constructed to shelter the troops. Baldwin, who had decided to begin construction with the labor of his own regiment, was shocked to find there were not enough tools available to begin clearing the abundant timber covering the mountain top. Baldwin stated that, if troops were to be stationed on Allegheny Summit during the winter, "measures cannot be too promptly or energetically taken and pressed for putting them into hospitable winter quarters."

General Jackson must have shared Baldwin's sentiments, because two days later, he issued orders assigning Captain William H. Tebbs of the 3rd Arkansas to command work details on the top of Allegheny. His command was divided into fifteen working parties, ten drawn from Camp Bartow and the other five from the 52nd Virginia. Tebbs's mission was to construct huts for winter quarters, storehouses, and hospital facilities. The work crews were to depart early in the morning, encamp at Allegheny Summit, and begin construction as rapidly as possible.[110]

The troops already stationed on the mountain had begun preparations to spend the winter there. Blankets arrived and the men were expecting to be issued overcoats any day. Work on fortifying the camp also began. About one hundred and fifty acres of timber was felled along the Green Bank road, where an artillery battery was to be emplaced. Captain John Miller's artillery company was anxiously awaiting the arrival of their cannon. In anticipation Miller drilled his troops daily in the operation of loading and firing, using two pieces of wooden cannon constructed especially for this purpose. One was placed on a tree stump while the other was mounted on an old two-wheeled wagon.[111]

While Confederates at Camp Bartow prepared for winter, small Union parties were sent out to discern the intentions of the enemy. On October 18, an unidentified Ohio scout had left Cheat Summit with "an old mountaineer for a guide, a Cherokee Indian, and two of my company, who are noted for their coolness and courage. They are all brave men." The scout noted, "The mountaineer has a price set on his head for guiding the Union

Army. He is a desperate man, and of course will die before he is taken."

Leaving at daylight, they traveled over the mountainous terrain and encamped for the night on the Greenbrier River. The scouts climbed atop a mountain, where they were able to clearly view Camp Bartow. After drawing a plan of the camp and its entrenchments, the Union men marched nearly ten miles farther to gain information from local residents. After completing their dangerous mission, they returned to camp on October 23.[112]

Meanwhile, on the morning of October 19, a heavy fog, described by one soldier as "the most dense fog I've ever seen," settled onto the Greenbrier Valley. Both sides sent reconnaissance parties out along the turnpike. The Federal detachment that descended Cheat Mountain consisted of several companies, while the Confederate party consisted of six men led by a Lieutenant Herbert. The Confederate scouts departed at 8 a.m., crossed the bridge beyond the outermost picket stand and proceeded toward the Union camp at Cheat Summit.

They had traveled but a short distance when they ran into the much larger Union force, which attempted to cut them off from their picket post. The Federals immediately opened fire, one ball passing through the shoulder of Herbert's coat. The Confederates quickly retreated to the first picket post at Slaven's Cabin where they awaited the enemy's approach. They were reinforced by the arrival of the other pickets under the command of Colonel Albert Rust. After a lengthy wait Rust decided to advance, moving his men forward to Trotter's stall, half a mile east of the bridge. Suddenly the enemy emerged from the thick fog. Gunfire erupted on both sides and the outnumbered Confederates again retreated, falling back in hopes of setting up an ambush closer to Camp Bartow.

One Confederate was wounded in the leg, but he was apparently the only casualty. The Union detachment advanced on the Arbogast house and set it ablaze before withdrawing to their camp. Rust then ordered the Confederate party forward, and they arrived just in time to extinguish the burning home. While the Confederate pickets claimed to have "killed or cripled some of the enemy," a Union report claimed they had not suffered a single casualty.[113]

When the fighting erupted, the 52nd was ordered to prepare to march down the mountain at a moment's notice. After waiting anxiously

Colonel Albert Rust, who commanded the 3rd Arkansas Infantry. Library of Congress, Prints & Photographs Division, Civil War Photographs [LC-USZ61-1094]

throughout the day, word arrived in the evening that the regiment's services would not be needed. The Pittsylvania Cavalry arrived to relieve the Churchville Cavalry, which left for Monterey on the following morning. According to John Lyon Hill, "Not many tears were shed as the somber peaks of old Cheat died away from our view." A scouting detachment from Camp Bartow fired on the Union pickets on Cheat Mountain.[114]

The Union soldiers encamped on Cheat Summit were treated to a strange sight on the morning of October 21. An elderly dark-skinned gentleman suddenly appeared outside the fort, vigorously waving a white handkerchief over his head. After being admitted inside the works, this civilian, apparently an escaped slave, revealed what little information he knew about the Confederate force at Camp Bartow. Union officers found a place for him in the commissary where, "comfortably situated," he cheerfully helped cook for the soldiers. Another soldier recorded that the escaped slave was Ben Summit, whom he described as "a quite intelligent contraband, who deserted from the rebel camp, and was installed as one of the servants at Head Quarters."

General Robert H. Milroy questioned Summit and learned that he was from Arkansas and had made good his escape after being sent by a Confederate officer to procure medicine. His father was a Cherokee, his mother an African American. He had no name other than "Ben," so Milroy christened him Benjamin Summit for the place where his freedom was secured.[115]

By mid-October, the Union troops were well-clothed—thanks to a persistent barrage of letters from Reynolds and his troops, pressure from Indiana Governor Oliver Morton, and Northern newspaper coverage of the ineptitude and corruptness of the Union quartermaster department in western Virginia during the fall of 1861.

The same could not be said of the Confederates at Camp Bartow, however. They suffered immensely from a lack of winter clothing and some of the troops did not even have blankets to protect them from the bitter elements. In an effort to combat the brutal weather, some of the Confederate soldiers built fires in their tents "by constructing a kind of furnace something like one flue of a fruit dry kiln. They make the tents right comfortable, and take much less wood than a fire out of doors."

On October 20, Lieutenant Colonel Michael G. Harman, quartermaster at Staunton, wrote to General Jackson concerning supplies for the troops at Camp Bartow, and noted corn and clothing were en route.[116]

On the same day, Jackson sent an urgent appeal to Harman, asking for his assistance in sending the 58th Virginia Infantry to Camp Bartow. Jackson feared that Union forces were attempting to outflank his position by moving up the Seneca road and was exasperated by the failure of the 58th to come to his aid. "We are here in the immediate presence of a largely superior force," Jackson informed Harman. "I cannot spare a man to go back, and yet this command, which could have rendered us so much service, and which I designed for this very duty, and which, had it moved, might have prevented, by its mere presence, this foray of the enemy, lingers in Staunton." Jackson noted that his position at Camp Bartow was necessary to deter any Union advance and urged Harman to induce the 58th to advance "for our country's sake."[117]

Two days later, Harman wrote to Confederate President Jefferson Davis in response to Jackson's complaint about the 58th Virginia. He informed Davis that a great number of the men were sick in the hospital and that the regiment's strength was down to about 400 effectives. Harman was apparently correct, as a citizen in Staunton wrote of the regiment, "The ranks are thin from sickness, etc."

Adjutant General Samuel Cooper had already ordered the 58th to advance immediately, but Harman asked Davis to post at least one regiment on the Hardy line to guard the Seneca road. He warned that a Union advance in force along this route could gain the rear of the Confederate forces at Camp Bartow and Allegheny Summit and take possession of the vast stores at Monterey. The 58th Virginia would depart from Staunton on October 24.[118]

Wednesday, October 23, was an extremely cold day with rain turning into snow. Still another Union force ventured forth from Cheat Summit on an expedition into the Greenbrier Valley. This detachment consisted of fifteen men from each company stationed on the mountain top, about five hundred altogether, led by Major Horace Bell. Bell, who had participated in William Walker's invasion of Nicaragua and served in the revolutionary command of Benito Juarez in Mexico, had done much of the mounted scouting for General Reynolds and now served as a volunteer aide to General Milroy.

One soldier observed, "It is his delight to command a select expedition and his dashing courage has led him into many bold adventures. He is about the most dashing looking soldier in these parts." The Union column was guided at first by a runaway slave, but they soon came across a Union supporter from the area (possibly the ever-present John Slayton) who volunteered to lead the expedition. About sunset the Union column drove in the twenty pickets manning the first Confederate picket post after a lively skirmish. Bell's men then headed to Arbogast's, and this time the Federals were successful in burning the home. The sound of gunfire was heard at Camp Bartow, and the light cast by the burning building could be clearly seen. Once more the Confederates formed for battle.[119]

The Union column marched rapidly to Green Bank by way of the road that turned off the pike near Slaven's Cabin. The few men guarding the town were quickly taken prisoner. Among them were Uriah Heavener and Joe Gum, who was described as "a real Virginia mountaineer—whose uniform consisted of linsy pants, an old quilt for a shawl, and an old hat with a strip of muslin tied around it." Gum, known as "a daring mountaineer cutthroat," had been returning home when he was captured, carrying his gun, some cartridges, and a sack of corn meal. When he pleaded for his life, the Union troops calmed his fears and even carried the meal home to his wife before marching him off to Cheat Summit. Heavener was later released.[120]

Bell's detachment also appropriated forty head of cattle and nine of the finest horses in the area, two of which belonged to a Mr. Hartman. One of the horses was a fine black stallion valued at $400. Major Bell paid a visit to the post office where he awoke the postmaster and called for the mail, a demand that surprised the man, who growled that midnight was a "curious time for the mail." Bell replied that "he could not help that but for him to hurry." The postmaster then "struck a light." Curiosity turned to panic when he saw Bell's uniform. "'My God,' said he. 'Our mail going to the North.' 'Yes,' said Bell 'and you with it so come along'." The Union detachment then returned to camp with their prisoners and livestock without encountering the enemy.

While a Union officer called it "one of the boldest affairs that was ever recorded," a Confederate sarcastically commented that the raid was "quite a brave bold act for a pack of thieves to plunder a neighborhood after night ocupied principally by helpless women." Union troops from the 25th Ohio and 9th Indiana scouted the area several times during the next few days in an attempt to capture Confederate mail, hoping to discover not only military plans, but also letters that might reveal which local citizens were friends and which were foes.[121]

About ten o'clock on the night of October 26, Company C of the 31st Virginia was ordered to report to headquarters, equipped and ready for action. Here they learned that their pickets had spotted a party of Union troops moving slowly up the Greenbrier River in an attempt to cut off the Confederate outposts. Rumors that the Union force was substantial and that an attack was imminent quickly spread through Camp Bartow. Company C moved forward and took positions around the Burner house. At midnight, Colonel Samuel V. Fulkerson led 200 men of the 37th Virginia and 200 from the 12th Georgia, under Lieutenant Dawson, to a house in the mountains midway between the camps, where a sizeable body of Federal soldiers were reportedly encamped. The Confederates left the road and ventured into the rugged forest, fighting their way through a maze of brush and briers. Arriving at their destination two hours before sunrise, they waited in silent anticipation for the first light of day. Fulkerson did not locate any Union troops, but he did find evidence they had been there and had constructed some crude breastworks. Their mission completed, Fulkerson led the soldiers back to camp.[122]

There was still a desire on the part of some Union officers to launch a new assault against the Confederates at Camp Bartow. Milroy and Major Bell rode down from the summit to meet with Generals Ebenezer Dumont and Joseph J. Reynolds to see if they were prepared for another offensive.

Born in Vevay, Indiana, on November 23, 1814, Ebenezer Dumont graduated from the University of Indiana, was admitted to the bar, and practiced law in his hometown. He served briefly in the legislature and also as treasurer of Vevay. During the Mexican War, he was captain and then lieutenant colonel of the 4th Indiana. Dumont became further involved in politics after the war, serving in the Indiana House of Representatives and as a presidential elector for Franklin Pierce.

On April 27, 1861, Dumont was selected as colonel of the 7th Indiana Volunteer Infantry, a ninety-day unit, and led his regiment into battle at Laurel Hill, Rich Mountain, and Corrick's Ford. He was promoted to brigadier general in September. Dumont was ready and willing to attack,

Union General Ebenezer Dumont. Library of Congress, Prints & Photographs Division, Brady-Handy Collection [LC-DIG-cwpbh-04563]

but according to one Indiana soldier, Reynolds "was very indifferent and seemed posesed of a very large amount of extreme prudence."[123]

General Robert Huston Milroy, in command of the Union brigade at Cheat Summit Fort, ordered another reconnaissance to probe the Confederate defenses around Camp Bartow. At least one company of the 24th, 25th, and 32nd Ohio regiments prepared to march, along with part of the 9th Indiana. This Federal force would be led by Major Horace Bell, whose successful raid on Green Bank a few days earlier had made him a hero, a status soon to be forgotten.

Captain Jeremiah Williams, who commanded Company C of the 25th Ohio, reported to Milroy at 9 p.m. on October 29 with forty-six of his troops. Williams had no lieutenants available for duty so he invited Lieutenant Charles Jones of Company B to accompany him. Jones agreed and brought along thirty volunteers from his company. Milroy instructed Williams to report to Major Bell at Gum Road, some eight miles from the Union camp. When Williams's men reached this point they found two pickets and an orderly who reported that Major John Milroy, the general's brother, had advanced to a barn three miles down the road and left orders for Williams to meet him there.

Williams also learned that gunfire had already erupted. Confederate pickets from Camp Bartow stationed at the outermost post had been fighting the bitter cold when suddenly a force of about thirty Union troops moved into sight. Momentarily frozen by the biting wind, the pickets quickly recovered and opened fire, checking the Union advance. The outnumbered sentinels retreated to the next post although a few were cut off in the woods.

Captain Williams sent six volunteers forward as an advance guard. These men proceeded about half a mile when they were met by Major Bell, who directed them to halt. Williams soon arrived and was ordered to continue to the barn. Bell ordered the advance guard to fall back into ranks, claiming that he had pickets stationed along the route and needed no additional "feelers." Williams moved his column forward and was alarmed to find that not a single picket guarded the road.

Williams's detachment reached the barn a short time later where they found Major Milroy with about 60 troops of the 9th Indiana. While waiting for Bell to arrive with the remainder of the companies, Williams inquired about the objective of the expedition. Milroy explained they were to drive in the Confederate pickets to within one and a half miles of Camp Bartow and to occupy the old hospital building until daylight. The Federals would then scout the mountains on each side of the Confederate camp before departing. Upon their withdrawal they were to drive away a herd of cattle supposed to be grazing between the hospital and Camp Bartow.

To his credit, Williams expressed grave reservations to Major Milroy about this dangerous expedition. Small-scale skirmishing had already erupted, blazing fires had been built for warmth, and the Confederates would certainly become aware of the Union force, which would be exposed to the more numerous enemy for twelve hours. In addition, the Federals were advancing blindly, with no knowledge of possible ambush sites, which the Confederates would undoubtedly utilize to full advantage. Despite Williams's objections, the expedition proceeded.

Bell arrived at about 1 a.m., and he immediately formed the Union companies and ordered their advance. The company from the 24th Ohio was deployed as skirmishers on the right; the detachment of the 9th Indiana advanced in the center; the companies of the 25th Ohio entered the woods on the left and advanced in line; and half of the detachment from the 32nd Ohio picketed the Gum Road while the other half was stationed behind a fence near the barn. As Bell gave his orders, Captain Williams noticed something suspicious about the major's conduct. Regardless, the Federal force began to advance on this bitterly cold morning.

Williams's detachment ran into immediate trouble, not with the enemy but with the dense underbrush, which made forward movement nearly impossible. The troops stumbled and fell in the mud and became stuck in the painful grasp of briers; few were able to proceed more than twenty paces. Williams halted the advance and moved his men by flank into a nearby field.

The Ohio soldiers proceeded through the field bordering the woods for a short distance until Williams decided to make another attempt. The captain headed into the woods alone but only managed to progress about ten yards before becoming stuck in a swamp.

After extricating himself from the quagmire, the captain turned to rejoin his company when he heard a voice from the woods cry out, "Halt!" Williams responded by asking the unseen soldier what he was doing in the woods, to which the man replied that he was a picket. Williams then asked to what regiment the soldier belonged. The response revealed that the hidden sentinel was a Confederate. Williams said no more and walked over to the other end of the line to consult with Lieutenant Jones. While they were conversing, they heard movement in the underbrush less than five yards away.

The other Union companies had found the march a little easier and were moving toward their objective. Captain Williams's detachment was in danger of being left behind. Williams rode forward and informed Bell of the difficulties his men had encountered. Bell directed him to post pickets across the field and to fall in the rear of the column with the rest of his men. The troops from the 25th Ohio moved onto the road where they formed into four ranks and proceeded, finding the march much less arduous. Company H of the 24th Ohio had also returned to the road and was in the advance of the column, followed by the company from the 9th Indiana. Suddenly a deadly volley was fired by Confederate troops who had set up an ambuscade in the bushes on the hillside. The balls ripped through the ranks of Company H. The troops of the 24th returned the fire before falling back a few yards to reload. They returned and were again greeted by another deadly volley. Four soldiers of the 24th were killed and seven wounded.

After returning the fire they again retreated, when Major Bell rode up and ordered them to advance once more. Captain Enoch Weller, who commanded the company, asked Bell if he was in charge of the expedition. When Bell replied in the affirmative, Weller stated, "I do not consider you in a fit condition to command; I think you are drunk, and I will not order my men to advance again."

This response infuriated Bell, who began cursing and berating Weller. The column had begun to bunch up and soon the order came through to "about face." The companies turned around and retreated at "double quick" for about fifty yards before the column halted and the wounded were moved through the ranks to the head of the column, now held by the two companies of the 25th Ohio.

While Lieutenant Jones ordered the troops of the 25th to begin their march back to Cheat Mountain Summit, Captain Williams rode back to consult with Major Bell, who was still arguing with Captain Weller, insisting that he lead his men on toward the Confederate camp. Williams interrupted and asked Bell to give the order for the expedition force to return to camp, but Bell refused, asserting he was going to advance again.

At this point Captain Williams told Bell that his companies were returning to the Union camp, with or without him. To this Bell again burst into a rage. Williams returned to his troops and led a detachment of six men to scout the road ahead to prevent another ambush. Three miles up the mountain, the wounded soldiers were placed in a building where they were treated by a physician. Williams remained with the casualties until a sufficient force arrived to guard the makeshift hospital. He then led his men back to Cheat Summit, while Bell continued his advance, noted by one Union man as "a singular mixture of tragedy and farce."

Major Bell carelessly rode on toward Camp Bartow with a small force, including a bugler who was mounted behind him on his horse. Bell led a charge that drove in a small Confederate detachment of forty men at the bridge, about 3 ½ miles from Camp Bartow. The Union troops then occupied the remains of the Arbogast home, where they built fires and erected a rail breastwork across the turnpike.

The Confederate pickets fled to Camp Bartow, many losing their hats along the way. A Confederate soldier who witnessed their return described them as "terrified." One of the shaken troops claimed to have been run over by about forty Union cavalry. The 37th Virginia, commanded by Colonel Samuel V. Fulkerson, quickly marched to confront the Union force, but by the time they reached the scene the Federals had withdrawn, leaving two dead Union soldiers and two others wounded.

Fulkerson observed, "They also carried off some wounded, whom we could trail for some distance by the blood." Only one or two Confederates were wounded in the day's skirmishes. Fulkerson also noted, "Their army is composed to a great extent of Dutch and Irish. I could see them killed with a great deal of bitter relish, if they were all Yankees. But the rascals make the poor ignorant foreigners do all the fighting that they can."[124]

When Captain Williams reached camp, he immediately wrote a report describing the expedition and the drunken condition of its leader, Major Horace Bell. After being signed by the captains of the companies involved, the report was given to General Robert H. Milroy. Several reports stated that Bell had misrepresented himself as a commissioned officer and was in fact not a soldier. Bell reportedly was seriously injured during the ill-advised reconnaissance and departed from Huttonsville on November 11. He reappeared later in the conflict, serving as a scout for various Union units. After the war, he moved to California, opened a law office, and wrote a book, *Reminiscences of a Ranger*, about life in southern California that is considered one of the most important works on the early days of the state.[125]

The spirits of Union forces in the region were uplifted during the last week of October by the arrival of a prominent visitor, Major Adam G. Slemmer, one of the Union's first heroes of the Civil War. In January of 1861, three months before the first guns were fired on Fort Sumter, Slemmer had been stationed at Barrancas Barracks in Pensacola, Florida. At the time he was a lieutenant commanding a United States artillery company. When Southern forces threatened his position, Lieutenant Slemmer moved his company offshore to Fort Pickens, where he was reinforced by about thirty seamen. The Navy Yard at Pensacola was soon occupied by enemy forces, but Slemmer refused to surrender Fort Pickens and was hailed in the North as a defiant hero. His arrival in western Virginia stirred the Union troops into a frenzy of washing and cleaning. The Union soldiers spent eight hours a day drilling, leaving barely enough time to eat. Slemmer inspected Reynolds's forces at Camp Elkwater and Camp Foster and declared them "the best-drilled of any in the service." Soon, however, Major Slemmer would be "confined to his bed due to severe sickness."[126]

In late October, the 13th Indiana embarked on one of the most difficult expeditions of the campaign. Word had reached General Joseph J. Reynolds that a small Union outpost in Lewis County known as Fort Pickens was threatened by a gathering militia force sympathetic to the Confederacy. Reynolds was determined to quell rebel supporters in Lewis, Upshur and Webster counties despite the difficulties in penetrating the region with a sizable force. The objective of the mission would not be revealed until several days into the expedition.

Soon the camp of the 13th Indiana was alive with activity as the men prepared for the expected ordeal. Provisions were drawn for three days and the soldiers hurriedly roasted and ground coffee. The men were delighted when new uniforms were issued, "consisting of dark blue muslin pants and sky blue kersey overcoat each one appearing as proud as a boy with his first boots." The collars were described as "tall enough to hide a little mans head." Governor Morton of Indiana had exerted strenuous efforts to supply clothing for the soldiers from his state, and the new uniforms would quickly become worn from hard use.[127]

At half past noon on October 28, the troops of the 13th Indiana, accompanied by Captain Robinson's company of Ohio cavalry from Fayette County, Ohio, departed Huttonsville and moved south. Sergeant James King saw the quartermaster and his assistants loading sacks of supplies upon the backs of mules and horses and realized that the march would be strenuous, "over hard roads or no roads at all."

The march, which began with the band playing "Yankee Doodle," led the regiment past their former camp at Elkwater, now occupied by the 2nd (West) Virginia Infantry. Camp Elkwater had been strongly fortified by its earlier inhabitants, the 15th Indiana, with extensive breastworks and a gate that commanded the Huttonsville-Huntersville Turnpike. Being without tents, the men spent the night in the bushes but were soon awakened by an early reveille.[128]

The 13th Indiana moved up the Elkwater Valley on October 29 and arrived at a former camp site where the men were permitted to rest and make coffee. During the brief respite, Dr. Swiger, a Union man who had previously practiced medicine in the region, arrived and offered his services as a guide. Swiger was no stranger to the regiment, having led a number of scouting expeditions. He immediately proved his worth by informing the Union commander that he was on the wrong road. Backtracking about a mile, the soldiers then crossed the river and struck an unfinished turnpike intended to link Summersville and Slaven's Cabin. The road, which was good in sections, led up a precipitous mountain.

After several miles of torturous climbing, the troops halted at Brady's Gate long enough for the butchers to kill and dress a few cattle. Another former camp site of the 13th, Brady's Gate was named for an old mountaineer who once lived here and kept a toll gate on the unfinished pike. Advancing two miles farther, the regiment set up camp upon the sides of a deep ravine. A fence, conveniently nearby, provided the troops with firewood and materials for shelter.

The sound of the bugle awakened the men, who formed in regular order in preparation for an address by the regimental commander, Colonel Jeremiah Sullivan. He informed them they were starting on a three or four day march to locate two rebel bands who were entrenched in the mountains.

Sullivan added, "Boys, I am anxious to have our own flag before going home," a declaration that met with enthusiastic assent from the men.

After proceeding less than a mile, the good section of road ended and the only way forward was by a rough, rocky path that meandered along the side of the mountain. After about five miles, the road again emerged, but it had been little used and was covered with briers. The Hoosiers were forced to make their way through an immense thicket of blackberries that stretched for miles. By the time they reached a clearing at three in the afternoon, hands and faces were bleeding and "many a blouse fringed which was unfringed before." The clearing revealed the first settlement seen during the day. The men of the 13th Indiana quickly prepared coffee, for no fires would be permitted after dark. Cattle were butchered in preparation for the next day's march before the troops rested for the night, comfortably hidden in the brush near the farm of Henry Moore.

On the morning of October 31, the troops marched for three miles along the side of the thickly wooded Point Mountain before beginning a precipitous descent of about the same distance, at which point flowed the swift and clear mountain water of the Back Fork of Elk River. During the crossing, a number of Hoosiers took a misstep and received a cold bath in the frigid waters. Proceeding up the bank for about half a mile, the regiment turned left and began ascending another mountain, cutting a path as they marched upward to the summit, where a clearing magically appeared. Upon reaching the far side of the clearing, the men began another descent and soon reached a small log house.[129]

While the advance elements of the regiment continued forward, following a beaten track along a small stream, Colonel Sullivan attempted to coax information from a young lady who resided in the home. To gain her confidence, Sullivan informed the woman that his regiment was from North Carolina, in the service of Jefferson Davis. This apparently swayed her, for she divulged that a number of secessionists had assembled to attack the Union Home Guards manning Fort Pickens.

A few moments later two scouts of the regiment rushed up to the rear of the column and informed their comrades that they had been fired upon from a house about one mile back. Sullivan immediately rode back and ordered Captain George W. Harrington to form his company and march double-quick to the scene of the incident. Dropping their knapsacks, Harrington's men started on a run for the house, which they quickly reached and searched to no avail. A young lady was present, but she claimed to have no knowledge of anyone with a gun. Suddenly a bullet whizzed by Captain Harrington's head. Turning toward the source of the shot, the Union soldiers could clearly see the smoke from the discharged weapon. Four men from the company started up the creek at the base of the mountain while others raced up the mountainside. Sergeant James King sighted the secessionist and fired at him, but the distance was too great and he made his escape.[130]

The rest of the company had also been busy, capturing five Confederate sympathizers, two of them on horseback. Their only weapons were knives and antiquated pistols. Also captured were fifty dollars in Confederate money and a pass through the lines at Camp Bartow. The Union guide, Dr. Swiger, recognized all of the prisoners and declared them dangerous characters, particularly William Bennett, who was the brother of the state auditor of Virginia.

The mounted prisoners were ordered to dismount and were replaced in the saddles by Captain Harrington and one of his lieutenants, who led their troops and prisoners back to rejoin the remainder of the regiment. Continuing to follow the stream, which they were forced to cross every five minutes, the men of the 13th Indiana marched until dusk when they reached the farmhouse of Watt Cool, "noted desperado and secessionist," where they were joined by three troopers of Ohio cavalry who had been sent forward to capture him. They missed snaring Cool but captured two of his sons and a mountaineer by the name of Boggs, who boasted to his captors that his business was killing Union soldiers. The men of the 13th encamped on Cool's farm, making liberal use of his beef, hay and wood.[131]

Early on the following morning, the first day of November, the expedition continued toward the forks of the Holly River. The morning's march was uneventful, but at about noon, while marching along the base of a steep mountain about fifteen miles from Fort Pickens, a heavy volley erupted into the column. Firing from a concealed position amongst the pine, laurel and underbrush, the secessionist force attempted to inflict heavy casualties on the exposed Union troops, but a steady advance by the men of the 13th Indiana soon put the rebels to flight. Colonel Jeremiah Sullivan climbed a nearby hill during the skirmish, and, "while standing near a tree, cheering on his men, a bullet passed within a few inches of him and striking the tree, scattered the bark and splinters about his head."

Three men of the regiment turned up missing, but only one man was wounded, and that a slight wound in the thigh of Private Alexander Doherty of Company I. Doherty was placed on a litter and transported to a nearby hut, which had been vacated by secessionist guerillas only moments before. The Indianans then settled in for the night after a march of eighteen miles and a fierce skirmish.[132]

The regiment continued on and reached Fort Pickens on November 3 after a difficult march through mud. There they found the small company of Union Home Guards entrenched and confident in their ability to withstand an attack. The Indiana regiment paused to partake of a light meal consisting of one cracker per soldier, although some managed to procure corn bread and molasses from the men guarding the Lewis County fort.

They departed from Fort Pickens on the following morning and crossed several mountains before marching along level ground through Centreville and Snatchburg and halting the day's march in Buckhannon. A number of troops from Company E "fell out and went into the houses" when they brought up the rear of the column during the following day's march.

On Tuesday, November 5, the 13th Indiana left Buckhannon, crossed the river and came to the pike two miles above the Middle Fork Bridge. They followed this road for about three miles before stopping for the night. On the following day they finally reached Beverly, where some of the men halted while others continued on to Camp Foster at Huttonsville with thirteen prisoners in tow. The soldiers of the 13th Indiana were exhausted by this taxing expedition that relieved the threat to the small Union force at Fort Pickens. The nine-day march, covering 186 miles, also made clear to Confederate guerillas that a sizeable enemy force could and would pursue them into their home regions.

A detachment of about forty troops from the 15th Indiana had departed from Huttonsville on November 3 to retrieve the wounded soldier of the 13th Indiana, who had been left in the care of a Union supporter in Webster County. The Federals struggled along a narrow mountain path, finally reaching the residence of a Mr. Buchannan, where the wounded soldier was recuperating, on November 5. Two days later, the Union detachment returned to Huttonsville without incident. Upon the return of the 13th Indiana, General Reynolds issued thanks to the men of the regiment.[133]

The Union commanders continued to send scouting parties down the Staunton pike to reconnoiter the Confederate defenses. According to a letter from a Virginia soldier to the *Richmond Dispatch*, one such expedition occurred on the night of November 2. Rain fell incessantly during the night and continued through most of the following day, at times turning to snow. Unable to build fires in the downpour, troops on both sides were forced to eat cold bread and beef. The Federal scouts attempted to burn down a house occupied by Confederate pickets but were driven off by brisk gunfire.

The Virginia soldier claimed that eight Yankees were killed and one captured, but these casualties were undoubtedly exaggerated. Regardless, as he wrote, "small parties come down every day or so, and 'take it out' in firing at a distance at our sentinels, and cursing them."[134]

On November 3, another such detachment of about eight Union soldiers opened fire on Confederate pickets at one in the afternoon. The Confederates returned the fire and forced the Union scouts to withdraw. The 12th Georgia was rushed forward from Camp Bartow to reinforce the pickets and soon reached the river, where they awaited news from a cavalry detachment that had been sent forward to investigate the disposition of the enemy. The cavalry soon returned to report that the Union force was insignificant, and the Georgia regiment returned to camp.

A short time later the Federals reappeared at the turn of the road near the Greenbrier bridge and again opened fire, this time from a distance of over half a mile. Despite the great distance, one ball nearly reached its mark, smashing into a rock within a few feet of a Georgia picket. These aggressive Union scouts, supposed by the Confederates to be drunk, remained in place for a few hours, cheering for Lincoln and "cursing the Damd Rebbles." One Confederate at Camp Bartow recorded that the firing "anoyed us no little in camp."[135]

CHAPTER FIVE
STONEWALL GETS HIS WAY

In early November 1861, Confederate Major General Thomas J. Jackson, newly appointed commander of the Valley District, began formulating plans for a winter offensive against Romney in western Virginia. Jackson requested and eventually received cooperation from General William W. Loring, whose Army of the Northwest was to be withdrawn from the passes over the Alleghenies. Loring agreed to the withdrawal due to the convincing arguments of Jackson and Confederate Secretary of War Judah P. Benjamin that there would be no Federal offensive along the turnpike during the winter of 1861-62. Again, the actions of the Army of the Northwest were dictated by their anticipation of Federal operations in the region.

Union occupation of Romney on October 26 alarmed Confederate supporters in the South Branch Valley and facilitated the planning of a counteroffensive that would determine future movements of the Confederate Army of the Northwest. Two heavily utilized corridors ran through the town. A north/south route crossed from Cumberland, Maryland, to Monterey, Virginia, where it intersected the Staunton-Parkersburg Turnpike. The Northwestern Turnpike led from Winchester through Romney and Grafton to Parkersburg. Romney's strategic importance is readily apparent in the fact that control of the town changed hands repeatedly during the Civil War.

On October 31, 1861, Confederate Lieutenant F. G. McCoy had informed General Henry R. Jackson that a Union force still held Romney and was "plundering considerably among the supplies of this valley." Another Confederate soldier, Joseph Vanmeter, wrote to Jackson on the first day of November concerning the problem and complained that the people of Hardy and Hampshire counties suffered due to an oversight that failed to include them in a military district.

Confederate supporters in these counties felt abandoned and watched helplessly as Federal troops raided the vast amounts of livestock, crops, grain, and supplies for their own army in western Virginia. Vanmeter complained that the enemy had moved unobstructed into the valley while men from the

Confederate General Thomas J. "Stonewall" Jackson. Library of Congress, Prints & Photographs Division, Civil War Photographs [LC-DIG-cwpb-07477]

area served in Confederate units elsewhere, and the supplies seized by the Federals were needed by the Confederates, who depended on this region to supply the Army of the Northwest. Vanmeter noted that the army's quartermaster had been purchasing and hauling corn from Moorefield to Monterey and asserted it would be much more economical to station a force of cavalry and artillery in the valley for the winter. He also pointed out a Federal force could march from Romney to Monterey, cutting off supplies for General Loring's army. Vanmeter closed with a prescient plea for a move to regain Romney: "I hope you will urge upon the generals the importance of sending a sufficient force to dislodge the enemy from Romney, if you concur in its importance."[136]

At least one high-ranking Confederate officer, General Thomas J. "Stonewall" Jackson, agreed with Vanmeter's assertion that Romney should be retaken. His insistence upon the cooperation of Loring's Army of the Northwest in his planned offensive would eventually lead to orders for a withdrawal of this army from the Greenbrier Valley. Jackson's plan for a winter offensive originated from his fear that the Federal force at Romney, which was being reinforced, was preparing to advance on Winchester to gain control of the town.

On the evening of November 4, Jackson arrived in Winchester to assume command of the newly formed Valley District, which encompassed the region between the Blue Ridge and Allegheny mountains. Jackson, who had been given little but his command, immediately beseeched Confederate Secretary of War Judah P. Benjamin for the use, if not control, of Loring's army. It is important to understand that Jackson did not have the authority to order Loring to comply with his request. This duty fell to Benjamin, who was reluctant to command Loring to join the expedition and sought instead to persuade him of the importance of the task.

Having frequently journeyed on the turnpike from Staunton to Cheat Mountain, Jackson knew that the road was nearly impassable in the winter months and proposed that the Army of the Northwest be withdrawn from the Allegheny passes. The western Virginia native firmly believed "that if the withdrawal of the Confederate forces from the Cheat Mountain region shall induce the enemy to advance on Staunton it will be his ruin, provided a sufficient available force is kept in this district in marching order." Jackson also reasoned a concentration of Loring's army at Winchester would produce greater opportunities for threatening northwestern Virginia than leaving it in a defensive posture in various posts guarding the passes.[137]

Reaction to Jackson's request by the Confederate secretary of war was swift and positive, leaving Loring with less than half his army. The Army of the Northwest at the end of October was comprised of five brigades, with a total of 11,700 soldiers present. On November 5, Confederate Adjutant General Samuel Cooper wrote Loring on behalf of Benjamin, ordering Brigadier General Daniel S. Donelson and two Tennessee regiments to reinforce General John B. Floyd, whose Army of the Kanawha was stationed in Greenbrier County. Another 6,000 troops were dispatched to Staunton to await orders, leaving Loring with 4,500 soldiers to hold the Monterey and Huntersville lines.[138]

Meanwhile, both sides continued to send out scouting and foraging parties. Early on the evening of November 7, the 37th Virginia Infantry under Colonel Samuel V. Fulkerson was ordered to march west from Camp Bartow along the pike to ambush a Union detachment reported to be posted astride the Slaven's Cabin road. The regiment crossed the bridge, then left the road and began crossing a ridge that led to Slaven's Cabin.

As they moved along the mountainside, Fulkerson began to doubt the wisdom of the mission. It was an extremely cold night, described by the commander as "the coldest of the season." He noted, "It had rained all the night before and was very muddy and the branches all flush. On going about two miles I found that the men's feet had become very wet by wading the branches and through the mud..." The shoes and clothing of his troops became soaked and quickly stiffened in the night air. The march was difficult, with the troops forced to battle briers and locust bushes as they passed along the eastern brow of the steep ridge.

Concerned that his men would get frostbite, Fulkerson detailed his quartermaster back to Camp Bartow to inform Colonel Edward Johnson of the impossibility of the assignment. Johnson, however, sent back an order that the regiment was to proceed with their mission. Fulkerson demurred and "at once determined to disobey the order." He promptly marched his men back to camp where he reported to Johnson. On the following afternoon, Confederate scouts reported that a detachment of Union troops had been sighted near Green Bank. Fulkerson's regiment was again ordered to Slaven's Cabin in an attempt to cut them off from Cheat Summit. Upon reaching their destination they found that "there had been but three men there for some days."The 37th Virginia returned to camp in the evening.[139]

Colonel Johnson was livid with Fulkerson's actions, and on November 9 he placed the colonel under arrest for disobeying orders. Fulkerson informed his sister, "Do not think that I am confined or guarded. I was only informed by an officer that I was in arrest and as a consequence I do not take command of the Regt. while in that situation."

Johnson explained: "The night referred to was not too cold for the enemy to be out. Our own pickets were out . . . If orders are to be thus disturbed discipline is at an end. If we are to turn our backs upon the enemy for such matter we ought to be arranging the terms of capitulation."

Fulkerson angrily replied that Johnson, "being himself in a warm house," could not have been "fully sensible of the severe character of the night." Fulkerson asked Colonel Seth Barton to investigate the incident as quickly as possible. He also requested a copy of the order assigning Johnson to command of the post, as well as the date of his commission, apparently to determine who was the senior officer.[140]

About 6 a.m. on the morning of November 11, a Union detachment opened fire on a Confederate picket post about three or four miles from Camp Bartow. After a few shots, the Federals ceased firing and began cursing at the pickets. Possibly a bit under the influence, they said they wanted to trade whiskey for tobacco. They continued the verbal assault for close to three hours before the pickets finally lost their patience and fired upon the Union scouts, quickly driving them off.

The troops of the 12th Georgia, who were preparing breakfast on this rainy morning, were ordered to reinforce the Confederate outpost. The regiment immediately marched to the pickets' relief, but the Federals had by this time retreated. According to one Georgian, "it was only a small force of Yankes that had come down on the side of the mountain and was fireing at our pickets all of a half a mile, their shots having no effect—only to annoy us was their intention." By the time the 12th Georgia returned to camp, the troops were "all wet and muddy up to our knees" and completely exhausted.[141]

Meanwhile, an incident occurred in Monterey that revealed the hostility caused by the guerrilla warfare in the region. Two captured Union men from Romney were brought to Monterey on November 11 and placed in the guardhouse. One of the men, whose name was Roarbaugh, had reportedly killed a Confederate cavalry scout of the Charlotte Cavalry and shot at least two others. His presence incited great excitement and indignation, and the guard was doubled to prevent an attempt to seize him. On the following night, an armed mob assembled about 11 p.m. and quickly overpowered the guard, seizing the sergeant, corporals and sentinels. They then dragged the prisoner a short distance from his cell to an open field in back of town and "cut him almost to pieces." His body was brought in on the following morning; one observer noted that the Union man "had some dozen stabs on his body and an awful cut on the back of his head."

Colonel George W. Hull, who commanded the post at Monterey, placed the sergeant and corporal on duty at the time under arrest and sought to apprehend the participants in the lynching. The troops at Monterey were summoned about midnight and roll was called to determine if any of the men were missing. Four cavalry troopers were called before a court of enquiry on the following morning but were immediately released when they provided an alibi for their whereabouts. One of the townspeople was arrested and placed in the guardhouse, but it is unclear whether anyone was brought to justice for the murder.[142]

On November 13, a Union scouting expedition was organized to descend into the Greenbrier Valley. The troops drew two days' rations and began their march about 9 a.m. About three miles from Camp Bartow the Union detachment divided into squads. Pickets were sent forward while parts of the 25th and 32nd Ohio and 9th Indiana regiments detoured east of the Confederate camp to round up cattle and horses from the Uriah Heavener farm near Green Bank.

Part of Company A, 32nd Ohio, remained at the forks of the road to guard against a Confederate flanking movement while the balance moved forward along the road toward Camp Bartow. This detachment was fired on about three o'clock in the afternoon by Confederate pickets guarding the Staunton pike. The Ohio boys returned fire with effect, sending a couple of the pickets "howling to the ground." The skirmish continued throughout the afternoon in front of Camp Bartow, but the gunfire was scattered and casualties were few. Private Robert G. Thompson of Company A, 32nd Ohio, was hit with buckshot in the right forefinger, while Lieutenant Abraham Norris had dirt knocked into his eyes.

The Union force advancing on Heavener's, led by Lieutenant Colonel William H. Blake of the 9th Indiana and Major Shelton Sturges of the 24th Ohio, surprised a detachment of the Pittsylvania Cavalry and captured Lieutenant John L. Hurt and a private. Nearby were Confederate officers Colonel George Hansbrough, a battalion commander, and Colonel James N. Ramsey, who commanded the 1st Georgia Infantry. When advance scouts of the Union detachment suddenly appeared, these officers were forced to make a rapid departure to avoid capture. Hansbrough and Ramsey rounded up about fifty members of the local militia under the command of Captain James C. Arbogast and set up an ambush along the Green Bank road, but they saw no more of the Union force.

At Camp Bartow, Colonel Johnson learned that part of the enemy was at Green Bank, but a persistent rain apparently prevented him from sending out a force from camp to cut off the Union detachment. Orders were sent to the 52nd Virginia, still stationed on Allegheny Summit, instructing this regiment to descend the mountain and find the enemy. The 52nd did not depart until early in the evening and was also unsuccessful in locating the Federals.[143]

Late in the afternoon the Union scouting party reached Heavener's, where they seized a few horses and some cattle. On their return from the raid, however, they released the colts due to their unruliness. The cattle were also granted their freedom when the Federals learned that Confederate patrols had blockaded the road. Avoiding the ambush, this Union detachment returned to the forks of the road about 3 a.m. where they rejoined the other elements of the expedition. With the exception of the two prisoners and a beehive, the Union troops returned to Cheat Summit with nothing to show for their exhausting efforts.

The rain fell all night and continued on the following day. Two companies of the 31st Virginia were ordered to Slaven's Cabin to reinforce the pickets at this post. Near Green Bank, an Indiana soldier who had become separated from his comrades the day before attempted to get directions to Cheat Summit from some of the local residents.

Hungry and hopelessly lost, Private Abel Chapman stopped at the home of a Mr. Carr and was cordially invited in for a bite to eat by two sons of Mr. Carr and a son of Captain Arbogast. The boys promised to guide Chapman to the Union fort, but only after a hearty dinner was consumed. During the meal one of the young men stealthily appropriated the rifle of the gullible soldier and soon Chapman was staring down the barrel of his own weapon. When the boys informed Chapman that he was to be taken to Camp Bartow, he refused, but he had little choice.

That evening the youngsters proudly marched into the Confederate camp with their prisoner, who was turned over to the Commandant of the Post. Chapman, a member of the 9th Indiana Infantry, was originally from Connecticut. At first he was a bit apprehensive as to his fate, but he was treated kindly and soon began to converse freely. Chapman was described by one Confederate as "a handsome, intelligent young man."[144]

By the morning of November 16, the rain had turned to snow, blanketing the Greenbrier Valley with nearly six inches of the fine white powder. In some places the snow drifted to a depth of two feet. Some of the Confederates at Camp Bartow awoke to discover that the drifting snow had invaded their tents and often their beds as well.

Enjoying a good night's sleep in the inhospitable climate of the mountains also proved difficult for the Union troops stationed at Cheat Summit, but the soldiers adjusted. Instead of straw, many of the men slept spoon fashion on cut pine tree tops or limbs with blankets and overcoats piled atop them. The day remained cold and the snow continued to fall. On November 17, a huge snowball fight lasting several hours erupted between the two Georgia regiments, with the 12th Georgia eventually emerging triumphant.[145]

Although Loring had apparently not determined whether he would commit the remainder of his army to the contemplated Confederate attack on Romney, on November 18 he issued orders for an eastward withdrawal of the Army of the Northwest. This order left three regiments and two battalions of infantry, a company of cavalry, and two batteries on Allegheny Summit. One regiment and part of a battery would remain stationed at Forks of Waters while another force of equal size was to hold a point between Petersburg and Monterey.

The remaining troops of the Army of the Northwest were to withdraw to Staunton. While the two brigades at Camp Bartow traversed the Staunton-Parkersburg Turnpike, the southern wing of Loring's army would march through Huntersville onto the Warm Springs road, reinforced by an artillery company to guard against a surprise attack on the withdrawing column.[146]

Confederate General William Wing Loring. Courtesy of Hunter Lesser

The early morning hours of November 19 were bitterly cold, but as the sun rose, its warm rays began melting away the snow. A squad of Federal troops again raided Heavener's and stole three horses belonging to Misters Heavener and Riley. Another detachment of Union troops was spied about seven hundred yards from the outermost Confederate outpost. One of the Confederate pickets fired his Enfield rifle and dropped a Union soldier, who was carried off by his comrades.

As the Confederates prepared to evacuate Camp Bartow, security had been tightened and very few visitors were permitted into camp. One Confederate, referring to the brief exchange of gunfire, remarked wryly that "one such group was turned away rather forcefully," though another recorded that the pickets had in fact "gallantly fired into a couple of poor horses." In any event, the troops were called out again at 2 a.m., but they were soon permitted to return to their blissful slumber.[147]

Loring still seemed opposed to participating in Jackson's offensive against Romney, possibly because it might place the remainder of his command under Jackson. He wrote Adjutant Cooper on November 19 to inquire if a further withdrawal of his troops was contemplated and to inform Cooper that "at least two regiments will be detached immediately from the Monterey line." Loring apparently objected to Jackson's plan, believing that leaving the passes over the Alleghenies was dangerous and would invite a Union advance on Staunton.

Cooper responded to Loring on the following day in an attempt to mollify the general's doubts. Cooper wrote that a previous order "was not intended to control your discretion in retaining such amount of force as you might find necessary for defensive purposes, &c., but only to make provision for such regiments as you might send from your command to Staunton." Cooper expressed hope that Loring could spare some of his men for service under Jackson or to reinforce the Confederate army at Manassas but left the decision to his discretion.[148]

General Thomas J. Jackson again wrote to Confederate Secretary of War Judah P. Benjamin on November 20, requesting that Loring's army reinforce him immediately for the proposed attack on Romney. In this dispatch Jackson finally revealed his plan of operation and the reasoning behind his desire for a winter offensive against the Union garrison stationed in the town. He believed that a concerted attack would enable the Confederacy to regain control of northwestern Virginia. Jackson conceded that a Confederate attack on Romney might convince Union General George B. McClellan that the Confederate force in his front had been weakened, thereby tempting him to attack. Should this occur, Jackson along with Loring's force would rush to reinforce the Confederate army at Manassas. After McClellan was repulsed, or if he failed to attack, Jackson proposed to lead his command "rapidly westward to the waters of the Monongahela and Little Kanawha…I deem it of very great importance that Northwestern Virginia be occupied by Confederate troops this winter." The Federals did not expect a winter offensive in this direction and abundant resources were available in the South Branch Valley to support his army. Jackson again disputed predictions that to counter the Confederate offensive, a Union force would move on Staunton, stating, "I am well satisfied that such a step would but make their destruction more certain."[149]

On November 21, preparations for the evacuation of Camp Bartow, scheduled to begin at 4 a.m. on the following day, continued. Extra baggage was shipped off in the morning to Allegheny Summit while many of the soldiers spent the day cooking and preparing rations for the march. Few regretted leaving "this miserable old place."

In the early morning hours of November 22, the eastward withdrawal of the two brigades of the Army of the Northwest stationed along the turnpike at Camp Bartow finally began. The drums beat reveille at 1:30 a.m. and the Confederate troops arose, struck their tents, and began loading the remaining baggage. The work dragged on until just before daylight, when the column began marching east up the rugged mountainside.

Most of the soldiers were disgruntled for one reason or another. Orders had arrived modifying the destination of some of the regiments. The members of the 12th Georgia Infantry were especially displeased when they learned that instead of marching on to Staunton as expected, they were to encamp on the top of Allegheny Mountain. Supplies and clothing which they expected to replace once they reached Staunton were to be burned or left behind. Compounding their anger was the fact that for nearly a month empty supply wagons had returned to the east when they could have been transferring the troops' possessions.

After a great deal of "swearing and foaming among the officers of small callibre," the 12th Georgia moved out, bringing up the rear of the brigade.

Watching the progress of the column, one Georgian described the sight as resembling a giant snake "slowly 'cralling on its belly'." A detachment of three hundred soldiers under Colonel Hansbrough served as rear guard of the column.

As the evacuation began, Hansbrough's detachment moved down the west branch of the Greenbrier River and set fire to the bridge "right in the face of the Yankees." Union pickets fired at Hansbrough's troops but quickly fled when they realized they were outnumbered. The Confederate rear guard then returned to camp, where they waited until the entire column was on the march before bringing up the rear. They finally reached Camp Allegheny shortly after dark.[150]

Although the withdrawal had been planned for over a month, it was very poorly executed and was compounded by foul weather. A drenching rain fell throughout the day, increasing the misery of the soldiers and turning the road into a quagmire of ankle-deep mud. The haste with which the withdrawal was conducted caused a myriad of problems. Tents and baggage were burned or thrown by the side of the road, broken wagons were abandoned, and many of the overworked, emaciated horses died in their harnesses and were tossed over the bank to clear the road. As the troops finally reached Allegheny Summit the rain turned to snow. Although detachments had been cutting timber for nearly a month, very few cabins had been erected. As a result, most of the soldiers were forced to pitch tents and suffered immensely in the bitter cold.

Secretary of War Benjamin issued an order to Loring on November 22 for the withdrawal of most of his cavalry. With the onset of winter, obtaining forage would be difficult. The horses were reportedly in "reduced condition" and were not expected to be needed in the mountains during the winter. A small mounted force from western Virginia would remain in the region while the remainder of the horses would be sent to eastern Virginia. Union forces also began withdrawing from the Cheat Mountain region for service elsewhere; on November 23, General Reynolds was directed to send three Ohio and two Indiana regiments to Covington, Kentucky.[151]

During the morning hours of November 23, Colonel William B. Taliaferro's brigade, which consisted of the 23rd, 37th, and 44th Virginia, the 1st Georgia and the 3rd Arkansas, packed up and marched east toward Staunton. The 44th Virginia halted at Crab Bottom where they set up camp for the winter. Some of the soldiers whose regiments would remain on the summit for the winter were angered by the departure of these troops. A member of the 31st Virginia was especially piqued by the withdrawal of the 44th, calling them "the grandest cowards in the army."

The Georgia soldiers were also upset over the arrangements. One wrote, "It seemed unjust to place us, some of whom are from a climate, almost tropical, upon these bleak, snow-clad mountains, and send Virginia troops, whose homes are here, and who are accustomed by life-long training, to the seasons here, into other portions of the service." A petition of protest was drawn up to be sent to the War Department, but as one Georgian noted, "I am afraid it is too late now."[152]

The departure of these regiments left Colonel Edward Johnson, who was placed in command of the forces remaining at Allegheny Summit, with the 9th Virginia Infantry Battalion, the 25th Virginia Infantry Battalion, the 31st Virginia Infantry Regiment, the 52nd Virginia Infantry Regiment, and the 12th Georgia Infantry Regiment. The Pittsylvania Cavalry would also spend the winter at Camp Allegheny, as would the Lynchburg Artillery and the 2nd Rockbridge Artillery. During the next four months, these troops would endure hardships that would never be forgotten by those fortunate enough to survive them.

The troops left on Allegheny Mountain had awakened on November 23 to discover that three inches of snow had blanketed the summit during the night. Private Joseph C. Snider, a member of the 31st Virginia, arose early. Snider, who had slept by the door of his tent, found himself covered with the snow, which continued to fall. Snider recorded in his diary that the air at this lofty elevation was "allmost as cold as Greenland." At about 10 a.m., Snider's company (Company C) was assigned three cabins, which had no roofs or chimney chinking. Despite the bitter cold the men immediately began dragging clapboards and rocks from the woods. Most of the completed cabins were given to the troops of the 12th Georgia Infantry.[153]

On the day following the Confederate withdrawal to Allegheny Summit, General Milroy led a 500-man detachment of the Union brigade stationed at Cheat Summit Fort east along the pike to reconnoiter toward the Confederate camp along the Greenbrier River. To their surprise they encountered no pickets and proceeded undisturbed to Camp Bartow, which they found deserted. While a small detachment moved up the mountain to scout the new Confederate camp, the remaining Union troops wandered about the battlefield investigating the fortifications and the damage inflicted on the buildings by the Union artillery during the fight of October 3.

They also found some letters left for them by some of their Confederate foes from Georgia. Private Thomas Prickett of the 9th Indiana related the contents in a note to a female friend: "They had left a letter addressed to our Reg. and one to the Gen commanding the thieves at Cheat Mountain. They gave our Reg particular fitz calling us cowards and everything that was cowardly. . . They invited us to come to Georgia and they would hang all of us."[154]

About 3 p.m., a report spread through Camp Allegheny that a Union force was advancing along the Staunton-Parkersburg Turnpike. Colonel Johnson ordered his troops into line of battle in preparation for an attack. Large scouting parties were sent both westward along the turnpike and down the mountainside through the woods. They discovered that Union troops were indeed nearby, but only a scouting party seeking to obtain information about Camp Allegheny. About half a dozen Federal cavalry had approached to within a mile of the camp and captured a sick soldier from the 1st Georgia who was recovering at a home along the turnpike before withdrawing. Upon reaching Camp Bartow, the Union detachment burned the houses and buildings formerly occupied by the Confederates, as well as the bridge over the Greenbrier, before returning to the summit of Cheat Mountain.[155]

The next few days were busy ones for the gray-clad troops on Allegheny Summit. The men placed roofs on their cabins, chinked and daubed the cracks to ward off the frigid winds, and built chimneys. The snow, which continued to fall, was nearly knee-deep and hampered the troops' efforts. One soldier described the cabins as "not very good, being built of logs with part stick and dirt chimneys." He added, "We have commenced to throw up some fortifications and are doing other things which should have been done before this time." Another noted that sixteen men were crowded into a hut 16' X 16', an average of one square foot per soldier. Despite these complaints, one soldier wrote, "we have very good times now. We can do almost as well here as we can at home."[156]

Camp Allegheny was constructed on the farm of John Yeager. His sugar maple grove of five hundred trees provided much of the lumber for the construction of cabins, and the back porch of the Yeager home was partitioned to house a commissary. Ten-year old Evelyn Yeager and her brother Mack were treated well by the Confederate soldiers and were often asked to make purchases for the men in return for some candy. They were also invited down to share the contents of packages soldiers received from home.[157]

The winter season would prove particularly hard on the soldiers from Georgia. "The mountains tower magnificently around us, and the snow upon their summits, glistens in the sunshine with a beauty to which our Southern eyes are unaccustomed," one Georgian wrote, recognizing that this scene might appeal to those with a strong love of nature. "But with us—such has been our experience—the mountains have lost their romance, and degenerated into huge heaps of vulgar earth; and we perceive nothing in the snow, but its power to chill and freeze."

Comparing a winter in the Rocky Mountains with one on Allegheny Summit, another soldier concluded, "I had much rather winter in the Rocky Mountains than on the summit of the Alleghany, from the simple fact that you could make your camps in the can[y]ons or deep ravines, so as to shelter yourself from the cold, howling winds which scarcely ever abates. The Rocky Mountains is the coldest place in America, and yet I had much rather winter in the Rocky Mountains than on the summit of the Alleghany Mountains of Virginia."[158]

While the Confederates prepared for winter, Federal troops in the Cheat Mountain region also constructed winter quarters. At Camp Foster these "fine looking Tabernacles" or "cow boxes," as they were described, consisted of "four sleeping rooms and four cook rooms to a company. The buildings are 18 X 22 and made out of logs, with a shed roof, bunks around the sides, no windows and no fireplaces. Very fine places for a 'soger' to freeze in."

Union troops on Cheat Summit also began building cabins. To aid the Federal soldiers in their efforts to combat the winter weather, a portable sawmill was shipped from Blandy's shops in Zanesville, Ohio, and operated at Cheat Summit under the direction of Lieutenant Albert Spaulding. A squad of men ran the machine, converting pine trees into boards used to build winter quarters.

A company of around one hundred men would construct four cabins measuring forty feet in length and twenty feet in width. Bunks lined the

Union fortifications and camp at Cheat Summit. West Virginia State Archives, Boyd B. Stutler Collection

walls on both sides, leaving just enough aisle space to squeeze through. At one end of the cabin, a huge fireplace with an attached chimney provided warmth to the men. By mid-November, more than one hundred log cabins had been built. The cabins were constructed in the form of a fort and "the outer side pierced for 'fox-holes'." Plenty of sweat was expended in building this "mountain city," as the lumber had to be carried more than half a mile.[159]

In this "mountain city," soldiers found various ways to keep themselves entertained. "In our cabin tonight, there are some twenty-five or thirty 'gay and festive cusses' assembled, and among them the 'fiddler,' who, after drawing his bow across a pine-knot for lack of rosin, leads off with *Dixie Land* and ten or fifteen 'pitch in,' with a vim worthy of a better cause," an Ohio soldier wrote to the Woodsfield, Ohio, *Spirit of Democracy*. The soldier continued his description of the camp scene:

> *In one corner a party of four are deeply immersed in that history, the author of which is Samuel Hart; in an opposite corner, two mathematical men are studying the intricacies of the chess board; while a third party are shuffling dominoes, manufactured from a pine shingle; still another party forms, and an old newspaper this time is the center of attraction. Some one is reading the news, and one by one the merry dancers gather around, the card players look up from their game and listen; the dominoes cease to rattle; chess-men lose their interest, and the reader and his unpretending little sheet becomes, for the time, the sole engrossing object. Even your humble servant, who has all this time been perched, like an owl, high over head in his bunk, arouses from his reverie, and lends an attentive ear. After listening till all are assured that "Washington is safe," and Lincoln still lives, the dancing commences, the games go on. What matter, though the wind does blow, and the snow fall, we are warm in our shanties, and will be merry.*[160]

The task of serving picket duty, dreaded by most soldiers, was especially unwelcome in the inhospitable climate on Allegheny Summit. "I with the rest of the company had to guard the ditches one night this week," wrote Charles Delay of the 25th Virginia, "and it rained and sleeted the whole night, and I never heared the wind blow harder in my life."

Another vivid description of the unenviable assignment was recorded by a member of the 12th Georgia, shortly after this regiment withdrew from Allegheny Mountain:

> *The howling winds and the sighing maple and snow clad mountains which greeted you in every direction, all seemed to help to make it one of the most miserable places that I ever saw. It was a very common thing to see them come in off of picket with icicles hanging to their heads. When we went out on picket we remained 24 hours before we returned to camp; again at each picket post we constructed a kind of Wigwam, by setting three forks together, letting them rest together at the top, and then spradling them out at the bottom so as to make them stand there, then we placed ten foot rails around it, similar to a potato hill, only leaving a hole for the potato, or Southern soldier to creep in at; dirt was then thrown up about a foot deep all around, leaving a little air hole at the top to keep the potato, or Southern soldiers from spoiling; from three to four generally stood at one of these posts, one being on watch all the time; a great many times while on picket if you did not keep walking, your feet would freeze to the ground.*[161]

On the night of November 25, Private Joseph C. Snider of the 31st Virginia, accompanied by three members of his company, started after dark to relieve a stand of pickets stationed east of camp. They "wandered about" until approximately 8 p.m. without finding the sentinels and they returned to camp. Snider and his companions were ordered to resume their search, which they did. They still had no luck in locating the picket stand and finally stopped at Varner's to spend the night.

Arising early, the four Confederates ate breakfast and renewed their search, without success, until about 9 a.m. when they again returned sheepishly to camp. Before they reached Camp Allegheny, they were met by the officer of the day who "took us back and posted us." The one consolation for the troops was that their cabin was finally completed enough to sleep in.[162]

On November 27, the temperature rose, changing snow into rain and turning the roads and ground into a giant sea of mud. The rain continued to fall for several days.

On November 29, Company B of the 31st Virginia returned from scouting duty in Pendleton County along with a company from Hardy County, which was a part of the 25th Virginia Infantry Battalion. Relief also came in the form of overcoats that arrived in the afternoon and were distributed to the grateful soldiers. Clothing supplies were often furnished by citizens who supported the Confederate cause, particularly women in the nearby counties.[163]

Although the Confederates were settling into their mountaintop quarters, no decision had been made about whether they would remain in the region for the winter. On November 24, Confederate Secretary of War Judah P. Benjamin had once again written to Loring seeking to induce him to order the remainder of the Army of the Northwest to withdraw from Allegheny Summit.

Benjamin was undoubtedly swayed by Jackson's belief that an offensive could lead to Confederate control of northwestern Virginia. In addition to the tempting prize of capturing the Federal garrison at Romney, Benjamin wrote that "a continuation of the movement westward, threatening the Cheat River Bridge and the depot at Grafton, would cause a general retreat of the whole forces of the enemy from the Greenbrier region to avoid being cut off from their supplies." The secretary of war agreed with Jackson's assertion that a Union force would not cross the Alleghenies to counter his offensive. "The enemy," Benjamin wrote, "would starve if dependent on supplies to be drawn from the valley or on supplies to be hauled across the mountains. It is quite late in the season for him to move over to Staunton and then go back across the mountain, and it appears to me that General Jackson is right in saying that his cross to Staunton would render his destruction more certain."

The Confederate secretary of war acknowledged General Robert E. Lee's concurrence with Loring's conviction that the passes had to be guarded through the winter and that the Army of the Northwest should remain in position. Yet he not so subtly reminded Loring of the difficulty in keeping open communications to the east during the winter months and of supplying and housing his army "in the severe climate of that mountain region." For now, the decision was left to Loring. If he objected to participating in the expedition, he could decline, but if he chose to participate, Benjamin requested immediate action and a plan of operations to be communicated to Jackson.[164]

Confederate Secretary of War Judah P. Benjamin. Library of Congress, Prints & Photographs Division [LC-DIG-ppmsca-05642]

General Loring's reluctance to join Jackson in his proposed attack on Romney was due in large part to his fear that leaving the passes over the Alleghenies unguarded would tempt the Federals in the Cheat Mountain region into advancing on Staunton.

Loring informed Benjamin on November 29 that his command would join Jackson but under certain conditions. He reiterated the danger in leaving open the passes to an enemy that still boasted considerable strength in the region. He added that huts for his soldiers were near completion, enabling them if necessary to winter in the mountains.

Loring also stated that due to the poor condition of the roads and the scarcity of transportation, it would take two to three weeks for his army to withdraw, and owing to the presence of enemy spies, he would not march eastward until sufficient transportation could be obtained to move his entire command. A large quantity of ammunition and over 250,000 pounds of subsistence stores were still at Huntersville and a reserve of ammunition was stored at Warm Springs. Loring sought time to remove these supplies to Staunton before the remainder of his army withdrew.[165]

The month of November closed with a bang on Allegheny Summit. Skyrocketing food prices led to increasing frustration. Butter cost forty cents a pound. Potatoes were four dollars a bushel; apples were five dollars a bushel. As one Confederate soldier noted, "It takes a month's wages to buy anything now."

The price of whiskey had also increased. Mr. Wilfong, who lived along the pike, had been doing a brisk business selling his product at the inflated price of five dollars a quart, taking advantage of the economic principles of supply and demand, with demand at this time exceptionally high. Colonel Johnson sent two companies "with charged bayonets" to Wilfong's to pour out the whiskey. This outraged the bored and cold Confederates who immediately went on a rip-roaring drunk, which spread through nearly the entire brigade. The guardhouse was quickly filled with drunken soldiers who had turned to fighting to fill the entertainment void. Two or three of the men were so unruly that they had to be "bucked and gagged." One Georgia soldier was "cut to pieces" with knives, dying instantly from the wounds. Perhaps the best explanation for these actions was recorded by a Georgian who wrote, "I don't believe there is a man living who can resist all the temptations with which he has to encounter in this horrible condition."[166]

In early December, Confederates stationed at Huntersville also got drunk, but with less deadly consequences. About five companies had been ordered here from Marlin's Bottom. The garrison was commanded by Captain D. W. Alexander, a cavalry officer from Tennessee. Alexander was described as a "kind man and a good fellow" but was considered incompetent by his men. He was cautious, constantly sending out scouting parties to ascertain the designs of the enemy, but made no preparations to greet them if they attempted to seize the town. Many of his men were without arms and discipline was sparse.

Shortly after their arrival at Huntersville, James N. Potts's company of the 31st Virginia encamped in an apple orchard. While cooking breakfast on a December morning, P. A. Tolley, a local citizen who was a friend to many of the men, rode into camp. After perfunctory greetings, Tolley informed the soldiers that he had a five-gallon keg of apple brandy in his wagon, a gift for the boys. Most of the overjoyed troops gave up their efforts at cooking, choosing instead a liquid breakfast.

One of the participants admitted, "for all practical purposes the company was drunk, but in the very best of humor. The whole day was given up for a general jollification." That evening several empty flour barrels were broken up and the staves extracted to be fashioned into paddles. After dividing into two companies, a game something like "prisoner's base" commenced. When a member of the opposing company was captured, he was "severely paddled." The game continued until ten o'clock that night, with "every one remaining in a good humor."[167]

By the first of December, General Loring had moved to Staunton to direct the movement of his troops. Four regiments were dispatched to Manassas by Adjutant General Samuel Cooper while Loring ordered two others stationed at Millboro to join him at Staunton.

Loring wrote to Secretary of War Benjamin at least twice on December 1, requesting guidance. In the second dispatch, Loring informed the secretary of war that he was "now pushing the sick and munitions to the rear as rapidly as possible, and have given orders to all trains to go back lightly loaded with grain and return with supplies, a large amount having collected both at Monterey and Huntersville." Loring's primary motivation for the second note appears to have been to express his displeasure with the order issued by Cooper moving four of his regiments without his authorization.

"It is proper to state," he wrote, "that, in consequence of movements made, in which I have been endeavoring to carry out your instructions, officers at a distance from my headquarters have been telegraphing without my authority to Richmond, the result of which has been a conflict of orders." Loring appears to have become sensitive to matters dealing with his command, but events had overtaken him.[168]

Adjutant Cooper wrote General Loring a brief and strained message on December 5 that undoubtedly reflected Secretary of War Judah P. Benjamin's frustration with Loring's intransigence. "The exigency requires the arrival of your entire command as rapidly as possible at Winchester," Cooper's dispatch read. On the following day, Benjamin informed Jackson that Loring would cooperate with him, although the secretary of war expressed regret that Loring's response would not be immediate.[169]

On December 7, Colonel Edward Johnson sent a dispatch to Loring's adjutant, Colonel C. L. Stevenson, in an attempt to ascertain whether his command would remain on Allegheny Mountain for the winter. His message reflected his belief that their presence on the summit was necessary:

> *If it is intended to abandon this position, under the impression that the enemy have left Cheat Mountain, or that if they have not, the roads & climate, &c., will prevent their making incursions into this country, a grave mistake has been committed. . . . If this post is abandoned there will be nothing to prevent their march to Staunton, and my opinion is that they will improve the opportunity thus offered them. Moreover, if they get possession here it will be difficult to dislodge them. Our own intrenchments will afford them shelter, and additional works will make this point very strong. The cavalry to be here will be, in my opinion, of no avail against the forces of the enemy. Little reliance can be placed in the cavalry I have thus far seen. Infantry and artillery I consider essential to hold this position.*[170]

While Johnson obviously thought his command should remain on Allegheny Summit, his superior had already put into effect plans to withdraw his force. Loring forwarded this dispatch to the Confederate adjutant general with a note expressing his belief that an infantry regiment, supplemented by a section of artillery, could sufficiently guard the Staunton-

Parkersburg Turnpike from the vicinity of Monterey. A cavalry detachment would remain at Huntersville, aided by militia from the Millboro area. The rest of the Army of the Northwest would join Jackson's expedition into northwestern Virginia.[171]

Despite the plan to withdraw, the Confederate soldiers continued working to improve Camp Allegheny in anticipation of a lengthy stay. According to Private B. F. Templeton of the 2nd Rockbridge Battery, by December 8, soldiers had built fourteen cabins and were building a 130' by 26' stable with a slate foundation that "had to be dug 7 ft to get a level base of 26 ft." They also were expected to construct a bath house, blacksmith shop, and hospital, as well as a road one-quarter mile in length.

A frustrated Templeton wrote, "all of this work says nothing for our defence and should we be attacked by the enemy we will have to fight without any protection. We could have got a good foundation for our stables in twenty steps of the presant one without digging more than 18 in but as the Capt says my men will make it he taking no hold of the picks."[172]

The fine weather continued. Four Confederate deserters were brought back from Pendleton and Hardy counties. However, five soldiers from Hansbrough's Battalion deserted from Camp Allegheny. One of these men, Private Jefferson Glenn, had provided a clue to his motive for deserting as well as his destination only a few days earlier. Coming off picket duty, Glenn remarked to Lieutenant David Poe that "we would get up one morning and find the Yankees around us." These deserters turned up later in the day at Cheat Summit Fort and were "marched off" toward Columbus on the same day.[173]

A train of wagons was sent from Camp Allegheny down to Heavener's to retrieve some hay for the emaciated horses. Another detachment of about twenty-five troops traveled down the pike until they reached Camp Bartow, where they surprised a Union detachment. The Federals fled across the Greenbrier before they stopped and looked back. The Confederate scouts feinted as though they were about to pursue across the river and the Yankees disappeared. The wagons were thus able to return safely from Green Bank in the evening.[174]

In the early morning hours of December 9, a detachment of about one hundred Confederates, accompanied by a few cavalry, moved down the turnpike to Camp Bartow. They crossed the Greenbrier and set up an ambush on the west side of the river. Within a short time a Union scouting party of about twenty-five soldiers appeared and the Confederate horsemen rode forward to greet them. They fired a round at the Federals before retreating in an attempt to draw them into the ambush. Despite strict orders to hold their fire, one Confederate sentinel fired prematurely, an act that quickly brought the Union soldiers to a halt. The concealed Confederates were forced into action, and they jumped out and fired a volley into the Federal scouts.

One Union soldier was killed, and a major was reportedly knocked from his horse. This officer was assisted in remounting and the Union detachment had to "make good use of their legs" until they were a safe distance away. Here they halted and began firing at the Confederates from long range before the return fire forced them to retreat toward Cheat Mountain Summit.

During the withdrawal, two Union soldiers from Indiana were captured. The prisoners were escorted to Camp Allegheny where, according to one Confederate, "The rush to see them was as great as if a monkey show had entered camp." The Confederates anticipated a larger encounter with the enemy. "We expect a fight here of before long," one wrote. "Captain Long came to our cabin a few minutes ago and told us to be in readiness." Word also spread through the camp that a withdrawal to either Winchester or Staunton was imminent. Skirmishing between the two sides was escalating, and Union soldiers also believed that a battle was imminent.

One of the Federals stationed atop Cheat Summit noted that "on the evening of the 9th our pickets was drived in by the rebels from the bridge so the 25th Ohio Reg was marched up on the mountain and expect to attack the cesesh soon."[175]

Also on December 9, General Thomas J. Jackson wrote once again to Benjamin, stressing the need for Loring to join him as quickly as possible. Jackson argued that Loring should leave the care of the supplies to a quartermaster and immediately move his command to Winchester. He also insisted that Loring's remaining cavalry should join his command. Jackson emphasized the importance of quick action, writing,

> *It does appear to me that the capture of General Kelley's army, including his munitions of war, would be of far more value to*

> *our Republic than General Loring's subsistence stores. If General Loring's entire command were here I would, with God's blessing, soon expect to see General Kelley's army, or a large portion of it, in our possession; but if General Loring is not here speedily my command may be a retreating instead of a victorious one, and the consequences of such a retreat may not have their disastrous effects limited to this district.*[176]

Following instructions from the War Department, on December 10 General Loring issued General Order No. 17, commanding the Army of the Northwest to withdraw from its present positions for "immediate service elsewhere." Johnson's brigade was to march directly to Staunton, while the force stationed at Huntersville would reach this point via Warm Springs and Buffalo Gap. Following the withdrawal of these troops, the Huntersville line was to be held by a cavalry detachment, supplemented by the Pocahontas Rescuers and Captain Jacob Currence's company from the 31st Virginia Infantry. The Monterey line would be guarded by Major George Jackson's cavalry company. In preparation for the move, commissary stores were packed and shipped east, and on December 11 the sick were moved to Monterey and Staunton. Loring added that it was important for the commanding officers to keep the troops together en route to Staunton to prevent them from wandering off into the countryside.[177]

On December 13, Loring slightly amended his previous orders, appointing Colonel John B. Baldwin of the 52nd Virginia to command the remaining troops on the Huntersville and Monterey lines. Baldwin was to take position with his regiment at McDowell and to "afford all protection possible to the adjacent country against depradating parties of the enemy."

In addition to his own regiment, Baldwin would oversee three militia companies and Major Jackson's cavalry. With this force Baldwin was to remain vigilant against any enemy movement in his front or on his flanks. Supplies for his command would be sent from Staunton to McDowell. Loring was obviously unaware of the events transpiring on Allegheny Summit on the day when he composed the orders.[178]

In Washington, Indiana Congressman Schuyler Colfax penned a letter on December 12 to General George B. McClellan, commander of the Army of the Potomac. Colfax urged McClellan to remove the 9th Indiana from Cheat Summit and transfer them to an area where an opportunity for battle would be more likely.

> *I have written you twice, urging the removal of the 9th Ind. Regiment and their brave General Milroy, all of them my constituents, from the Cheat Mountain Summit to some region where they can see an enemy. Any new Regiment can guard a mountain half as well as they who have proved by repeated conflicts, with the enemy, some of them under your own eye, in Western Virginia, that they are amongst the bravest of the brave. Others in Indiana, more influential than myself, have procured the release of other Indiana Regiments from that Siberia to fields where they will have a chance to fight instead of to freeze. All of Milroy's Brigade (3 Ohio Regiments) have left; but the Regiment of my neighbors, despite their brilliant service hitherto, seemed specially marked for a winter of inactivity at the Summit.*
>
> *Gen. Cameron promised to intercede with you specially for their removal. My colleague, Col. McKee Dunn, asked it urgently last fall. I now beg and earnestly entreat you to relieve them. They would prefer going to South Carolina; but Kentucky, Washington, any where where they can fight rebels, they long for. In Kentucky, with our new Ind. Regiments, you could easily make an Indiana Brigade for Gen. Milroy.*
>
> *Resp. Yr*
> *SCHUYLER COLFAX*[179]

In light of the events of the following day, this letter must be considered one of the more ironic footnotes of the campaign.

CHAPTER SIX
THE BATTLE OF ALLEGHENY MOUNTAIN

By the first of December, most of Union General Joseph J. Reynolds's army had been transferred to other theaters of operations, and General Robert H. Milroy commanded the newly formed Cheat Mountain District. Posts were manned at Beverly, Huttonsville, Elkwater, and Cheat Summit. Only four regiments were to remain under Milroy's command, but two other regiments, both from Indiana, had not departed from the region before Milroy decided in early December to attack the Confederates entrenched on the summit of Allegheny Mountain. The resultant Battle of Allegheny Mountain[180] was the bloodiest contest of 1861 in western Virginia, but its most significant consequence was the decision by Confederate authorities to retain troops at Camp Allegheny for the winter of 1861-62.[181]

Robert Huston Milroy was born on June 11, 1816, near Salem, Indiana, the son of General Samuel Milroy, one of the early settlers of the Indiana Territory and a member of the first constitutional convention of Indiana. The younger Milroy graduated from Norwich University, a military institution in Vermont, in 1843 and served as a company commander during the Mexican War. Milroy practiced law in Delphi, Indiana, and was later appointed circuit court judge. In 1854, he moved to Rensselaer to practice law. Even before Lincoln's inauguration, Milroy issued a call for organization of a volunteer company in Rensselaer and the surrounding area. He was later chosen colonel of a regiment, the 9th Indiana Volunteers.

When Milroy received his commission, the citizens of Jasper County, Indiana, presented him with a four-year-old dapper gray horse named, fittingly enough, Jasper. His company was the first mustered into service from northern Indiana and "the first regiment from Indiana to tread upon the sacred soil of Virginia." After its three-months service was up, the 9th Indiana returned to the Hoosier State and was mustered out of service at Indianapolis on July 30. Milroy went home for four days but then left to recruit a new regiment. On September 3, 1861, he was promoted to brigadier general, and two weeks later he arrived with his new regiment, also designated the 9th Indiana, at Camp Elkwater.[182]

Union General Robert H. Milroy, who commanded Union forces in the Cheat Mountain region during the winter of 1861-62. Library of Congress, Prints & Photographs Division, Civil War Photographs [LC-DIG-cwpbh-03184]

Although no orders or dispatches have been uncovered that would reveal the motivation for the attack on Camp Allegheny, other sources provide possible explanations. Several contemporary accounts claimed the attack resulted from information furnished to General Milroy by Confederate deserters. On December 9 five Confederate soldiers who had deserted from Camp Allegheny turned up in Beverly, where they apparently convinced Milroy that the soldiers of the Confederate brigade stationed on Allegheny Summit were demoralized and ready to surrender. Several sources asserted this information led Milroy to the decision to attack. Others believed the general was determined to strike a blow before his former regiment, the 9th Indiana, departed the region for service in Kentucky. Other possible explanations are that, with his new command, Milroy may have become overly aggressive, or he may possibly have been searching for glory and promotion. Although it appears Milroy initiated the plan to assault Camp Allegheny, his superior, General Joseph J. Reynolds, was aware of the planned attack and allowed 300 volunteers of the 13th Indiana to join him even though they were scheduled to depart from Beverly at any time.[183]

It is also unclear what Milroy planned to do if the attack succeeded. A brief occupation of Camp Allegheny would necessarily have followed, but an advance upon Staunton would have been unrealistic considering the size of Milroy's force and the difficulty of hauling provisions from Webster (four miles south of Grafton, Taylor County), due to the inclement weather and the atrocious condition of the turnpike. In fact, the Union force would most likely have returned to Randolph County whether victorious or defeated. In any event it appears that Milroy had been persuaded that the Confederates on Allegheny Mountain could be defeated and captured without much resistance.

Unfortunately, the Federal advance was handicapped from its inception. A large number of officers from each of the regiments were home on leave; their absence would prove critical. Also, despite the difficulty of the mission, participation was apparently voluntary for several of the regiments due to the large number of sick soldiers. Colonel Thomas Ford of the 32nd Ohio was too ill to make the journey and only 130 men from this regiment, led by Captain William Hamilton, joined Milroy's expedition. The operation was conducted by a total force of just less than 1,900 Union troops, which also included 250 of the 2nd (West) Virginia, 300 of the 13th Indiana, 400 of the 25th Ohio, and 700 men from the 9th Indiana. In addition, thirty of Captain James R. Bracken's Indiana cavalry rode along, joined by seventy-five men of Captain Silas F. Rigby's artillery company who "went along without arms, expecting to take possession of the enemy's guns when captured." The failure to attempt the admittedly difficult task of hauling artillery up Allegheny Mountain would also prove crucial to the outcome of the impending battle.[184]

The 13th Indiana and the 32nd Ohio departed from Beverly on December 11, while the 2nd (West) Virginia marched north from Elkwater. All three regiments camped for the night at Huttonsville in an open field. Stationed here was the 25th Ohio, a part of which would join the march on Allegheny Summit. One soldier wrote, "The air being very frosty, and having brought no blankets with us, we found our quarters rather cool." Joined by this regiment, the Union force was awakened by a bugler at 4 a.m. on December 12 and continued the grueling march along a nearly impassable road of deep mud covered with a crust of ice "which broke at every step."

The Federals arrived before noon at Cheat Summit, where they joined the 9th Indiana, which held the mountain post. While resting on the summit the soldiers of the 9th Indiana invited the men of the 13th Indiana to their quarters and gave them coffee and a nice dinner before the march resumed. The road between Cheat Summit and Camp Bartow was well macadamized and in fine condition, making the descent a relatively easy one. The journey down Cheat Mountain was quick and the Union troops were surprised to note that the substantial bridge across the Cheat River had not been destroyed. The war had brought much destruction to this lovely place, however; all the houses within three miles of the second bridge, where several skirmishes had taken place, were in ashes. As they approached Camp Bartow the soldiers learned that earlier in the afternoon a sharp skirmish had been fought by their advance guard.[185]

Although the Confederates stationed on the top of Allegheny Mountain were preparing to withdraw to the east, Colonel Edward Johnson continued to send patrols out along the turnpike as far as Slaven's Cabin, several miles west of their former camp on the Greenbrier River. On December 10, Johnson had issued orders to Major John D. Ross, instructing him to take a scouting party down the mountain and "ambuscade a point on the turnpike just beyond Camp Bartow, on Greenbrier." When Ross asked for volunteers, 106 troops of the 52nd Virginia stepped forward to accompany him on the expedition. Ross's command moved down the Green Bank road on December 11, guided by a local citizen familiar with the mountainous terrain. The Confederate troops marched two miles beyond Camp Bartow, halting at the narrows near Slaven's Cabin. They lay in ambush until nightfall before withdrawing into a recess in the mountains, where they spent the night. Early on the morning of December 12 they returned to the narrows and took position in a dense pine thicket along the Greenbrier River opposite the turnpike.[186]

About 3 p.m. fourteen Union soldiers were seen walking east along the pike unaware of the trap into which they were about to fall. These troops, from Company G of the 9th Indiana, were the advance guard of a Union detachment sent by General Milroy to construct a makeshift bridge over the Greenbrier and to hold Camp Bartow until his main force arrived in the evening.

Among them was Private Joseph R. T. Gordon, a 17 year-old native of Versailles, Indiana, who was known to his comrades as Josie. Gordon, a frail youngster, had enlisted in August of 1861, much to the chagrin of his father, Major Jonathan W. Gordon. The elder Gordon, a veteran of the Mexican War, had enlisted as a private with the three months' volunteers of the 9th Indiana and served with distinction in the early campaign in western Virginia. He was the first to reach the body of Confederate General Robert S. Garnett following his death at Corrick's Ford and was described as being "always prominent among the most daring for fearless but not foolish courage."

Major Gordon had already been authorized to recruit a battalion and was engaged in that endeavor when he learned of Josie's decision. He expressed his forebodings to his son in a stern, but touching letter:[187]

> *You ought to have done such an act from motives so noble, that, although you felt it to be an act of disobedience to me, it was still an act of obedience to God; and, therefore, not a deed to be ashamed of. If you have no such motive, then you have entered upon a hard service, and made a great sacrifice, which never can be*

> *rewarded either here or hereafter. Without such a motive I should feel that you had thrown away your life. If you were actuated by such a motive, and will still remember it, and keep alive in your heart, during the service, it may not be so bad. True, you have made a greater sacrifice than you can now understand or appreciate—even though you get through the service most favorably. But I do not wish to discourage you. With your unformed habits of life and thought, I fear you have just about thrown yourself away. I hope, however, that I may be mistaken, and that you will labor to convince me that I am. All that I can do to contribute to your success and happiness in the army shall be done. You need not be afraid that I shall interfere with your choice. I would sooner have died than you should have thrown away the season of life wherein men alone can prepare themselves for greatness and usefulness; but would now sooner die than you should do any unworthy act in your new vocation to bring reproach upon yourself or family. . . Remember your mother. Do not forget that you were among her last thoughts—one of the objects of her last solicitude and prayer . . . Write to me often; obey your officers, and die sooner than be a calf or a coward.*

His father need not have worried. The younger Gordon was a favorite of General Milroy and had quickly proved his mettle as a scout "familiar with every mountain, valley, and path around the enemy camp." Josie Gordon was at the head of the advance guard that approached the old Confederate camp.

As they came directly opposite the concealed Confederates, the trap was sprung and a deadly volley was unleashed upon them. Two Union soldiers, Private Jonathan G. Porter of Company G and Charles Rhoades of the regimental band, were killed. Five others were wounded in the hail of bullets that forced the survivors to flee or take cover. One Union solder ran behind a tree and began firing his pistol at the Confederates. Major Ross demanded that the man surrender, but he refused, deciding instead to make a break for freedom. It was a fatal mistake, for as he began to flee, he was shot and killed by Lieutenant William H. Wooddell.[188]

Even before the Union soldier's body hit the ground, Ross's men saw two Indiana companies rushing toward their position, drawn by the gunfire.

The Confederates estimated their number at one thousand, certainly an exaggeration, but they were now outnumbered by the Federals. The major ordered a retreat, but before they could get away the Federals poured "a fusillade of bullets" into them. The Confederates escaped, suffering only one minor casualty, and made their way back up the mountain to Camp Allegheny. "We retreated for about five miles and our party got divided," one Confederate wrote, "and we stationed our party at the forks of the road for to keep the yankees from getting possession of the road untill the ballance of our party came up and we held that point untill dark and our party did not come to us and we being about 10 miles from camp we thought it would be best for us to make our escape."

Reinforcements from the Confederate camp, part of the 12th Georgia, marched six miles down the mountain before returning to camp; on their way back they were overtaken by Major Ross's scouting party. Conflict was narrowly averted when Ross ascertained they were about to fire on their own men. An order to send two companies of the 31st Virginia to the relief of Ross's detachment was rescinded when it was learned they were safe.[189]

Meanwhile, the remainder of the Union column neared the old Confederate camp. The bridge over the Greenbrier River about two miles from Camp Bartow had been destroyed by the Confederates when they withdrew in late November, so the Federals crossed one at a time on a temporary foot bridge constructed from unburned timbers by the advance guard and marched on to Camp Bartow.

It was 10 p.m. before the entire force arrived. The men rested; many had been marching for two days. They also perused items left behind by the former occupants of Camp Bartow when they withdrew to Camp Allegheny. The discovery of Sunday School papers, a Bible, and a number of letters to the Confederate soldiers from their families and friends was to one Federal officer "a revelation which went far to modify my feelings towards the Southern people."

The night was extremely cold and General Milroy allowed his men to build fires for warmth. This was yet another mistake, because the Confederates at Camp Allegheny could clearly see the flames from the lofty heights.[190]

Milroy called his officers to a council of war where he revealed his ambitious plans for the assault. He had decided to divide his command, seeking simultaneously to attack the Confederates in force from two directions. Milroy was to lead the men of the 25th and 32nd Ohio and the 13th Indiana, as well as the cavalry and artillery detachments, up the mountain on the turnpike before moving into the woods to outflank the Confederate positions guarding the road. The second column, led by Colonel Gideon C. Moody, consisted of the 2nd (West) Virginia and the 9th Indiana. They were to follow the road toward Green Bank before turning up the mountainside along a small trail that led to the Confederate camp. Milroy ordered Moody's column to depart immediately, giving it an hour's head start. As this column departed, they passed by the bodies of the Indiana boys who had been killed in the skirmish, which had been placed by the side of the road and covered with blankets.[191]

Despite Milroy's forethought, the tardiness of this force in reaching Camp Allegheny would prove crucial to the outcome of the battle. Colonel Moody's column was led by John Slayton, whom Confederate Colonel Edward Johnson referred to as a "noted guide and traitor." Slayton led the Union troops along the Green Bank road for several miles before turning east up a steep mountain trail that led to the left of the Confederate camp. The march was slowed by fallen timber and the terrible road over which they traveled.

One story told after the war claimed the Federals found a few barrels of apple cider at the Nottingham homestead and refused to march one step farther until the cider was emptied into their mouths and canteens.

Whatever the reason, the tardiness of Moody's column would prove to be the decisive factor in the battle's outcome. Slayton led the men off the trail into the woods between Salisbury and Buffalo runs, along a ridge later known as Slaten's Ridge. Hours later, as they moved through the wooded wilderness, they could hear the sounds of battle, but although they grimly picked up the pace, the obstacles placed by the Confederates and by nature continued to impede their advance toward Camp Allegheny.[192]

While Moody's column moved along the Green Bank road and up the mountain, the other Union column marched up the turnpike, some of the exhausted soldiers reportedly walking while asleep.

About 4 a.m. on the morning of December 13, the Confederate pickets posted on the road about two miles from the summit heard the approach of the Union force, still half a mile away. When the advance guard of the column neared their position, they opened fire, killing one Union soldier before falling back to form another trap along the pike about two or three hundred yards from camp. Here they were reinforced by Hawkins's and Blandford's companies from the 12th Georgia, sent by Colonel Johnson to help slow the attack expected from that direction. The ambush caused some troops of the 25th Ohio to stampede in fear through the ranks. Captain Thomas M. Kirkpatrick of Company E, 13th Indiana, wrote, "They came back on double quick, running over our company, knocking some of the boys off of the pike, causing them to lose their caps. They run over Barney Busby and Moses Gordon bruising the boys considerably." The sight of a dead comrade being carried down the mountain further demoralized the men. According to 2nd Lieutenant Ulysses Westbrook of the 32nd Ohio, it was "a mistake to carry the corpse through our lines, and I think it had a dampening effect upon all." His captain, William Hamilton, moved to the front of the ranks and requested that the officers leading the advance of the column devise another method of handling casualties.[193]

The soldiers at Camp Allegheny, numbering about 1,200 effectives, were aroused by the roll of drums as soon as the picket firing was heard. The Yeager family, whose home was in the middle of the camp, was awakened by the loud conversations of the Confederate soldiers. Looking out they could see officers riding back and forth on their horses issuing orders. One large body of troops was forming in line of battle and marching in the direction of the crest north of the old log church, while another camped just south of the turnpike double-quicked to the trenches and breastworks behind the Yeager homestead.[194]

Johnson had ordered the 31st Virginia along with Hansbrough's and Reger's battalions to occupy the high ground on the right of the camp, where a blockade had been constructed from fallen timber. Hansbrough's Battalion took positions in the woods above and to the right of the blockade. The 31st Virginia was stationed several hundred yards to the left of Hansbrough and did not link up with his battalion, possibly due to the difficulty in moving through the fallen timber in the dark. Hansbrough sent pickets to scout in both directions along the turnpike. Soon after dawn one of three scouts from Company A, who had ventured down into the woods and returned, reported that the enemy was advancing "six files deep at the double quick."[195]

About a mile from the camp the Union force had left the pike and marched up the mountainside toward the top of the ridge. Colonel James A. Jones was placed in command of the attacking force while Milroy remained with the reserve, which consisted of Bracken's cavalry and Company K of the 13th Indiana. Jones's orders were to advance to a position at the crest of the mountain and wait quietly until the sounds of gunfire on the other side of the camp marked the arrival of Moody's column, at which time both forces were to attack.

When the advance led by Lieutenant Isaiah B. McDonald reached the top of the mountain, however, they were fired on by Confederate pickets hidden behind fallen timber. The volley for the most part clipped the leaves and limbs from overhanging trees, although one Union soldier was killed.

While some of the Confederates were captured, the others retreated and Jones ordered a rapid advance to prevent the pickets from revealing the presence of the Union force. Pursuit ended at the woodline overlooking Camp Allegheny when the Federals observed about 300 Confederates advancing toward them in line of battle. Lieutenant McDonald, aide-de-camp to General Joseph J. Reynolds, reacted quickly, halting the advance company (from the 13th Indiana) and ordering it to deploy into line to the right to receive the attack. Colonel Jones arrived and ordered the 25th Ohio to the right of the Indiana company and two other companies of the 13th Indiana to its left, while the 32nd Ohio was posted on the far left.[196]

The Confederate force advancing toward the Union line was Hansbrough's Battalion, a unit considerably smaller in numbers than the force they were attacking. Hansbrough confidently determined "to feel the enemy with a view to test his strength and temper" and thus led his men forward to within 150 yards of the rapidly forming Federal line. Recognizing the danger he confronted, Hansbrough sent a messenger to request the immediate assistance of the 31st Virginia. He ordered his battalion to fire and fall back to meet the 31st. Lieutenant David Poe of Company A commanded his men to fire at about the same time a Federal officer immediately opposite them gave his troops the order.

Captain William Hamilton of the 32nd Ohio recalled, "We could scarcely see them for the smoke, but the noise was terrific. The men became confused and began to waver. I myself felt uncomfortable until a bracing thought came to me, and I called out 'Steady, boys, keep steady. Remember we are making half this noise'." After delivering a volley, Hansbrough's Battalion retreated through the fallen timber "like chased partridges" as a hail of return fire flew around them. Hansbrough's Battalion was pursued by the entire Union line, running, as one soldier wrote, "as hard as we could go pouring the shot into their backs and yelling at the top of our voices."[197]

As soon as he heard the battle open, Johnson ordered Blandford's and Hawkins's companies (K and A, respectively) of the 12th Georgia to march immediately to the support of the force engaged on the right, which was outnumbered three to one. In the meantime, the 31st Virginia, led by Major Francis Boykin, was advancing toward the fight. The Confederate regiment became disorganized and scattered, however, as they made their way through the felled timber. According to one soldier, they were "mixed up every way & could do nothing. Our Major beged us to form almost with tears in his eyes."

Finally a few men of Company C formed in two ranks and began marching toward the enemy, and the rest of the companies, led by Company A, formed and moved forward slightly to the left of Company C; these troops were somewhat protected by a point of the blockade. The troops of Hansbrough's Battalion fell in behind the 31st and followed them toward the advancing Union troops. Their commander stood atop a log and sought to rally his men for a charge, but as he descended, he was hit in the right thigh by a pistol ball. Lieutenant W. J. Robinson carried Hansbrough from the battlefield and the leaders of the Confederate companies ordered their men into line. Although Hansbrough's men had slowed the Union advance, their withdrawal had given the enemy possession of the crest of the ridge, forcing the Confederates to advance uphill.[198]

The Union attack was halted and the two forces held their positions for about two hours of fierce fighting. The small Confederate force was joined by a detachment of three companies of the 25th Virginia, who charged up the steep hill, making their way through the blockade. When they breathlessly reached the top, a volley from part of the 31st Virginia mistakenly raked one of the companies, but they quickly recovered. The Augusta Rifles, a company of the 25th Virginia, were soon in the thick of the fight. As they reached the blockade of fallen timber, Captain R. D. Lilley was informed that Private John T. Wood of his company had been wounded. Lilley found Wood, who had been struck in the arm with a ball. Private Wood was willing to continue to fight, so his captain moved on. Wood moved behind a tree, took deliberate aim, fired, and immediately reloaded. As he started to lower his gun again, Wood was mortally wounded by two balls which entered the upper part of his forehead, knocking him unconscious behind a log. A passing Confederate yelled to Captain Lilley, "one of your men is shot." Amid the hailstorm of balls Lilley rushed to Wood, who was in the last throes of life. He remained only a moment before turning Wood onto his face and then turning once again to meet the enemy.[199]

The Confederate cavalry would do little fighting on this day. As the Union force attacked on the right, 2nd Lieutenant C. E. Dabney, who commanded the Pittsylvania Cavalry, rode to Johnson's headquarters to "get orders." Not surprisingly, Dabney was unable to locate the colonel, who was in the thick of the fighting. The lieutenant returned and, finding the fire too heavy for his troopers, led them back to the entrenchments where they dismounted and held the reins of their nervous mounts.[200]

The intensity of the battle produced incidents that stirred feelings of guilt. Private Joseph Snider of the 31st Virginia recalled: "I fought for more than an hour within a rod of two of my cabin mates who lay weltering in their blood. It being a very cold frosty morning a fog kept raising from their blood. One of them asked for help." He sought to console himself for his failure to attend to his brother-in-arms by noting that "it would have been certain death to have undertaken to have carried him off besides we had all that we could do to hold our ground." Another member of the regiment wrote,

> *Many of our men suffered intensely of their wounds before they died. Many were groaning from extreme pain, with the cold, clammy sweat of death upon their brows. I hope never to witness such scenes again. I was particularly distressed with the sad fate of my esteemed friend John Nutter, 1st Sgt. in Co. C. Early in the battle I saw him raise his hands and fall. I hurried to his side, but saw he must soon die. I spoke to him, but he could only raise his eyes and smile a faint recognition. He asked very faintly for water, but I could not tarry long with him. We hurried over here in the wild excitement of the hour, and left him there to die alone. When I saw him again he was dead.*[201]

The old log church, which was used by the Confederates as a commissary and storage building, was the scene of some of the bitterest struggles of the day. West of the church, the Confederates were forced to fall back south of the turnpike. As one rebel retreated, he looked back and saw a Union soldier kick down the door of his cabin, emerging a second later eating a piece of bread. The intrusion of his home angered the Confederate, who fired a shot at the offender. Although he did not know whether his aim was true, after the tide of battle turned in favor of the Confederates, he found a wounded Union soldier outside his cabin. The Yankee begged for water and his foe readily assented, even placing his knapsack under the soldier's head for support.[202]

Although none of the soldiers had occasion to take a nap, as some had at the Battle of Greenbrier River, several paused in the midst of the heated battle to examine wounds. The right leg of Lieutenant David Poe was knocked out from under him by a bullet. Poe "sat on a stump and made an examination and found that I was not much disabled." Early in the contest, Private William Batts of the Muckalee Guards (Company A, 12th Georgia) was slightly wounded in the calf about six inches above the ankle. "I had my gun at ready when the ball struck me," Batts wrote his sister shortly after the battle. "I continued loading and fireing for some time after until my leg began to hurt me very much when I sat down by the trunk of tree and examined the wound and saw that it was very slight."[203]

The tide of battle slowly turned in favor of the Confederates. In addition to the sixty men of the 25th Virginia, the 31st Virginia, Hansbrough's Battalion, and the two companies from the 12th Georgia, Johnson ordered three more companies of the latter regiment forward to reinforce the thin Confederate line.

Although still outnumbered, the Confederates were able to press the offensive. Joining the fight was Colonel Edward Johnson, who had lost his sword and picked up "an immense war club." Johnson urged his troops to advance "with curses and vaporings of his club. Woe unto the man who showed any signs of palpitation of the heart or weakness in the knees; he was made to realize that there was danger in the rear as well as in the front."

Lieutenant Lewis S. Thompson led Company A of the 31st Virginia forward, yelling above the din of the battle, "Come on, my brave boys, follow me." With a yell, his company charged forward, but at that instant

Thompson was shot through the body and fell mortally wounded. Colonel Johnson was only a few feet away when Thompson was hit and later made mention of his "good conduct."[204]

The dead lieutenant's brother, Captain William P. Thompson, was cut off from his men and found himself surrounded by the enemy. He somehow managed to escape unscathed although his clothing was pierced numerous times by bullets. When he learned of his brother's fate, Captain Thompson retrieved his body and returned to the Confederate camp. On the way he came across Private Martin V. B. Collins, a member of Hansbrough's Battalion who had been shot just over the left eye. Thompson told Collins that the "damn Yankees" had killed his brother and that he was going to kill one of them in return. The mortally wounded Collins replied to the captain, "they have killed me too, you must kill one for me." Collins somehow found the strength to walk back to his cabin where he was discovered sitting on his bed "all in a gore of blood" by a shocked tent-mate, who frankly informed Collins that he must prepare to meet his God. Remarkably, Collins lived for three days before finally losing his battle with death.[205]

The Confederates concentrated their force on the enemy left, and with an earsplitting yell charged forward and broke this part of the line, manned by the 25th Ohio and 13th Indiana. Jacob Pinick of the 32nd Ohio conceded, "It came our turn to run now and away we went and the sesech after us."

A number of Confederate soldiers were killed or wounded in this charge. Joseph Snider of the 31st Virginia was hit in the left hand by buckshot, and his gun was dented by a bullet, making it nearly impossible to reload. Snider simply picked up another and continued to fight. The pursuit halted at the woodline, where the Federals found protection among the fallen timber.

A large number of Union troops, largely from the 25th Ohio, fled to the rear. Although some were rallied by officers and returned to renew the struggle, many others would not fight again on this day. After about half an hour heavy fire was heard in the Confederate rear, forcing them to withdraw to confront this new threat.[206]

The force that suddenly appeared on the battlefield was the Union reserve, ordered forward by Milroy in an attempt to forestall defeat long enough for the attack on the second front to open. At the point where the column had ventured from the pike, Milroy had called Captain George W. Harrington aside and ordered him to detach twelve soldiers to advance up the road about a quarter of a mile to prevent the approach of the enemy along the pike. Sergeant James M. King headed this detachment and carried out the order, but as soon as they reached their position, the deafening eruption of gunfire was heard, sounding to King "as though the mountain had been sent asunder."

Two reserve companies of the 13th under Captains Henry A. Johnson and Wharton R. Clinton were ordered forward by General Milroy, and King's detachment was ordered to charge up the hill with them. The Union soldiers double-quicked for about a quarter of a mile, when they spotted a Confederate battery, which due to the nature of its position they had little chance of capturing. They continued forward up the steep mountain slope until they reached the summit, where they suddenly found themselves confronted by a body of Confederates less than three hundred yards from another artillery battery. These were the two companies of Georgians ordered to the front when fighting broke out.[207]

The appearance of the Union reserve to the rear of their left flank stunned the Confederates. Private William Batts, who had been examining his wound, grabbed his gun and "was ordered back to my hut, by this time there was the greatest confusion that I ever saw in my life we were scattered to and fro and the yankes in the same condition. There were Yankes on bowth sides of us." The Union soldiers fired a volley that killed Captain William Mollohan of Hansbrough's Battalion, which had withdrawn to confront the threat in their rear. Lieutenant David Poe ordered his men to file to the right, which put them over a hill out of range. As the Union reserve observed this maneuver they fired another volley, which mowed down witch hazel bushes in front of the battalion.

The Federals were in danger of being cut off, however, and the time had come to make their escape. Union Sergeant James King wrote, "We were entirely surrounded by our enemies, and there was nothing left but to fight our way out. The order was given to fire and then as soon as loaded again 'Charge bayonets' which was done with a will, and we cut our way out taking 13 of the enemy prisoners, and losing but one of our men. Once out of their ambush we took to the cover and commenced fighting again."[208]

The Union reserve force had been forced to retreat, but not all of the troops made it from the field of battle. Sergeant King wrote, "It was then that I learned that my friend Jones was killed. He fell dead struck in the mouth by a rifle ball which passed through his head only saying as he fell 'Rush on Boys' and was dead and died as a soldier need not fear to die. . . It saddened my heart to think that we could not bring his body off of the field. But what matters is where the body lies. We left him to be buried by his enemies upon the spot where he had so gallantly fallen, and there his body lies upon the bleak and cold summit of the Alleghenies."[209]

The appearance of the Union reserve had enabled the Union force on the Confederate right to regain lost ground. Efforts by the Confederates to turn the left flank of the Union line and get in their rear were repelled by detachments from the 13th Indiana and the 32nd Ohio with such vigor that the Confederates were ordered to fall back to the cabins with the Federals in hot pursuit. "We again rallied," recorded one Union soldier, "and the command was given to charge bayonet and into them we went on double quick but they couldn't stand it so away they went and we give it to them in their backs. This was the last charge we made and the hardest of the fight."[210]

Despite the terror of battle, there were humorous incidents during the struggle in the midst of the Confederate camp. Private Daniel Murphy of Company E of Reger's Battalion observed a number of Union soldiers taking cover behind a tent. As soon as one exposed himself, Murphy fired, killing a Federal soldier. Almost immediately the Confederates charged forward. Murphy ran up to the dead soldier and began rummaging through his haversack looking for something to eat. Instead he found a flask of French brandy. Crying out, "here's to you boys," Murphy took a hearty swig and then hid the bottle inside his tent before rejoining his company. Asked after the fighting concluded why he had hidden the brandy, Murphy wittily replied, "Oh I was afraid I might get struck and get the bottle broken." One Confederate complained that the enemy "got into our winter quarters and ate our bread and bragged about it."

There were humorous incidents on the Union side of the battlefield as well. After driving a squad of Confederates into the protection of a log cabin, the troops of the 25th Ohio paused momentarily to give three cheers to the astonished defenders. One soldier of this regiment, pressing forward with a blood-curdling yell, was shot in the heel. The Union soldier, described as a large, stout man, threw his gun into the air and began limping across the front of the Federal line, hollering, "Oh Lordie! oh Lordie!" The troops of the 25th responded with hysterical laughter and yelled that he should take care or he might get shot in the other heel. One man from the regiment wrote that the incident "did a great deal to restore self-possession among the men."

In the midst of battle, George Harvey of the 32nd Ohio complained to his captain that there was something wrong with his weapon. After the battle, it was discovered that a Confederate minie ball had entered the barrel of his gun at the same time that he had fired at the enemy. "Bullets have been known to meet in the air," noted his captain, "but this is the only instance I ever heard of during the war in which a bullet of one side crossed the line to hide itself in the barrel of the other."[211]

Colonel Johnson rallied his troops and a determined counterattack ensued, aided by the decision to utilize two artillery pieces upon the exposed Union force. One final charge was made by the considerably reduced Federal force in an attempt to bring off the dead and wounded, but when the Confederates once again counterattacked, the remaining Union troops withdrew from the field of battle, concluding the attack by Jones's column on the right of Camp Allegheny. As Captain William Hamilton of the 32nd Ohio was withdrawing with his company, his brother Robert stopped him and said, "William, I am hit." A ball had gone through his leg, breaking a bone and cutting a nerve. Because he was unable to walk, he was placed on a blanket and carried down the hill.[212]

The Confederate artillery attempted in vain to reach the Union reserve, which had returned to the bend of the pike. Although no one was injured by the shelling, the guns were bothersome, and General Milroy determined to silence them with an attack from the road. All but sixteen of Bracken's cavalry had marched to the rear of the Union force on the Confederate right in an attempt to rally the troops who had withdrawn from the field of battle and taken refuge under a spur of the ridge. Led by Milroy, the remainder of the Union cavalry charged up the road in an effort to capture the Confederate guns, which were on a bank fifteen feet above the road, protected by felled trees. The shot and shell flew harmlessly over the heads of the Union cavalry as they approached the battery.

The Confederate gunners, angered because they were unable to further lower their cannon, began heaving stones and shot over the bank at the Federals. Milroy soon realized that the fire of his troopers' carbines was unable to penetrate the thick protective timber. In addition, the Confederate force that had nearly cut off his reserves was marching rapidly to meet this new threat. The general ordered his cavalry to retreat to the bend in the road, which was reached safely despite a heavy volley from the Confederates and continued shelling from the cannons.[213]

As the fighting died out on the Confederate right, gunfire erupted on the left signaling the arrival of Moody's force. The head of the column surprised five members of the lone Confederate picket post on this side of the camp just as the men were preparing breakfast. The Union advance fired a volley which killed two and wounded two, but the gunfire gave notice of their presence to the entrenched Confederates. As the Union column neared Camp Allegheny, Confederate Captain Pierce B. Anderson, who commanded the Lee Battery, observed the advancing soldiers but concluded that they were pickets withdrawing in the face of the enemy.

Captain Anderson, a resident of Tullahoma, Tennessee, was the son of Joseph Anderson, one of Tennessee's first United States senators who later served as Comptroller of the United States Treasury. The younger Anderson attended West Point with Robert E. Lee and Jefferson Davis but after three years was permanently disabled from a bayonet wound in his right wrist. He studied law in Winchester, Virginia, and was soon employed as an attorney in Tennessee. He married the granddaughter of General Andrew Lewis and served a number of terms in the Tennessee state legislature. When the Mexican War erupted, Anderson raised a volunteer company that served in every important engagement from Vera Cruz to Chapultepec, where seventeen of his company were killed. His bravery during this battle led the company's survivors to present Anderson with a sword, on which was inscribed, "Never mind it boys. Company, forward! Remember, you are Tennesseans!" In 1861 he was elected major of the 1st Tennessee regiment, but as only one officer of this rank was permitted per regiment, he resigned and formed an artillery company in Lynchburg, Virginia, known as the Lynchburg Lee Battery. Anderson's battery had participated in the Battle of Rich Mountain. After the disastrous defeat, the Lee Battery withdrew to the Confederate camp on the Greenbrier River.[214]

Captain Anderson ordered the Confederates to hold their fire as he rode to the trenches and waved the oncoming troops forward. It was a tragic mistake. The attacking Federals fired a deadly volley that fatally wounded the captain, knocking him from his horse. The Union troops, forced to advance in column rather than in line, poured forward almost to the breastworks, but the entrenched Confederates responded with a furious volley that halted the attack and pushed the Federals back into the fallen timber and brush.

Captain Thomas Madden of the 9th Indiana, who jumped atop a log and urged his men forward, was seriously wounded in the charge while several others were killed, including Private Josie Gordon, described "as brave as a lion and as gentle as a lamb." Colonel Gideon C. Moody ordered his men to fall flat on their faces behind the fallen timber, a directive which undoubtedly saved many lives. From here they maintained a blistering fire on both the concealed infantry and the exposed artillery, although one Union soldier believed that "if our officers had done their duty and given the order to charge bayonet I am satisfied that we could have taken them." The remainder of the struggle on this side of Camp Allegheny was fought, as one Confederate noted, "almost after the Indian method."[215]

After the repulse of the initial charge, the Union troops had little chance of forcing the Confederates from their entrenched positions. According to Private John Chilcott of the 9th Indiana, "The balls flew around as thick as hail." Lieutenant Alfred Sickman of the 2nd (West) Virginia picked up the rifle of a fallen member of the 9th and fired without effect at a Confederate officer. After reloading, he raised up to fire again but was struck in the chest by a minie ball and died within minutes. Standing to fire exposed one's self, as Thomas Prickett learned: "Sy Mackey was standing about 6 ft from me he was shot through the head his brains were scattered all around." He noted Mackey never knew what hit him.

Prickett was angered by the failure of one of his comrades to participate in the dangerous contest and "finally told him to get up and go to work." The soldier complied and a short time later was shot through the arm, which later had to be amputated at the shoulder. Actually, Union casualties in this attack were much lower than those suffered during the struggle on the Confederate right. While the battle on the other side of the camp had been fought for the most part across open ground, the fighting on the side nearest the Confederate trenches turned into a stalemate after the initial assault was repelled.[216]

The Yeager home was situated in the midst of the torrid fighting, and hot lead pelted down on the roof of the home like hailstones. John Yeager's wife and daughter were both ill at the time and had been unable to flee the battlefield before the Union attack began. During a lull in the fighting, Colonel John Baldwin of the 52nd Virginia sent a detachment to carry the women to the relative safety of his own cabin, where they remained through the night of December 13. At the height of the battle the rest of the family was ordered out of the house and moved down Block Run.[217]

Meanwhile, Captain John Miller's battery kept up a blistering fire on the Union column that had taken cover in the fallen timber. Miller's guns were just to the rear of the trenches and the grape shot flew dangerously close to the Confederate infantry. At the height of the struggle on the Confederate left, an infantry captain who was seated on the ground against a tree facing the guns angrily yelled, "Capt. Miller you are firing too low!" Miller denied this was true, but the officer insisted, "But you are, sir, you are grazing my men every time, and grazing my head too." Miller replied, "Oh, you are mistaken, Capt., my shot go to the enemy." The exasperated captain finally threatened to report Miller if he did not raise his guns, but Miller stubbornly refused. After the battle the infantry captain had Miller arraigned for reckless shooting.

Miller, a Presbyterian minister, eloquently defended himself and was subsequently acquitted of the charge. Following the conclusion of the trial, Captain Miller walked over the ground where the infantry had been stationed during the battle. He was shocked to find that "a large grape-shot had entered the middle of the tree not three inches above the head of the captain who was sitting there!" Miller immediately ordered the tree cut down and burned. A post-war newspaper account noted, "Some persons might have felt a doubt as to whether this was the best course, but the redoubtable Captain had no doubt whatever."[218]

The Union soldiers, unable to advance, directed much of their attention on Miller's gunners, who had "no protection whatever." One member of the 2nd Rockbridge Artillery was killed and five were wounded, one mortally. Bullets repeatedly struck the artillery pieces. Even the entrenched defenders were not entirely safe from Union fire. One Confederate complained,

"They could lay behind the brush and logs concealed and when we would raise up to look for them they would fire at us they knew exactly where to look for us in the trenches."

Captain James Deshler, Johnson's acting assistant adjutant general, was shot through both thighs. Deshler refused to leave the field of battle and lay exposed to heavy fire for several hours.[219]

The Federal force finally withdrew from their positions near the Confederate trenches, bringing the long and hard-fought battle to a close. An ambulance and a number of soldiers sent to retrieve the Union dead and wounded was fired on by the 2nd Rockbridge Artillery but was permitted to carry out its gruesome task after raising a white flag. Not all of the Union casualties, however, were recovered.[220]

As the fighting concluded, the difficult task of treating the wounded and burying the dead began. Amputations were performed on at least seven of the Confederate wounded. In the Union camp, however, the surgeon declined to operate, claiming, "I would rather not take such a responsibility."

Robert Hamilton of the 32nd Ohio was forced to undertake a lengthy journey home to Zanesville, Ohio, to consult with a local surgeon. A Dr. Hildreth removed Hamilton's wounded leg, but the delay in amputating led to his death less than a week later. Several dead and severely wounded Union soldiers were left on the battlefield; the fate of others was even more macabre. As the head of Moody's defeated force neared the old Confederate camp at the base of the mountain they were horrified to discover a drove of hogs feeding on the bodies of the dead soldiers which had been left behind. The hogs had literally defaced the dead. The men fired a volley at the swine and began the grim task of interring the remains. To Ambrose Bierce of the 9th Indiana, "it seemed that they had moved and tossed their coverings off."

The inability of the Union force to bury their fallen comrades would lead to a similar horror story a few days after the battle. Lieutenant David Poe of Hansbrough's Battalion was in charge of a picket post near the camp. A heavy snow had fallen and Poe noticed a large number of fox tracks just off the path. Curious, he followed the tracks until he found some of the animals eating the bodies of Union troops killed during the battle. Their fellow soldiers had covered the bodies with bark and leaves, but the foxes had managed to find them. Poe returned to camp, and after Colonel Johnson learned of the incident, he sent a detachment out to bury them properly.[221]

The troops of the 44th Virginia, stationed near Hightown, had heard the sound of cannon fire and responded by undertaking a forced march toward the battlefield. On the way, they met several wounded Confederates who conveyed pessimistic reports about the outcome of the fight. Upon reaching Camp Allegheny, however, the regiment was met with cheers by their victorious comrades. Private William Price viewed the dead Union soldiers without sympathy, with one exception: "One of the dead was a youth apparently about 19 years of age. He had evidently fought very desperately as his face was blackened with smoke through which drops of perspiration had traced their lines. He had fallen near a large maple tree behind which he had taken shelter and was one of the most advanced & exposed points at any time occupied by the enemy. A ball had pierced his heart, and he was lying upon his gory couch of leaves, as composedly as if he had died without a sigh or a groan . . . He must have been a good and brave boy, the recipient of his mother's tenderest love . . ."[222]

Confederate parties worked late into the night combing the battlefield for dead and wounded. About 11 p.m. a detachment of Company C of the 31st Virginia stumbled across several wounded Union soldiers. "One of them was shot through the thigh and groin," one member of the detachment recalled, "He swore frightfully, cursing every one of us and saying, 'If our men had all fought like I did, there would have been none of you left to tell the tale.' Poor fellow he died the next day."[223]

At Camp Allegheny, officers and soldiers alike expected another assault on their position. "We slept upon our arms last night, in the open air, near our ditches," wrote one Confederate soldier, "thinking probably the Yankees would renew the attack last night or this morning." Trenches were extended, brush was trimmed and burned, roads blockaded, and additional fortifications were built to protect the batteries; in one soldier's words, "making all necessary preparations to receive the 'yankees' when they come to pay us another visit. They met with a warm reception when they visited us on the 13th inst., and if they should come again I imagine that they will met with a warmer one."[224]

In the days following the battle many of the Confederates sat down to write their loved ones, who would undoubtedly soon hear of the battle and would be worried as to their fate. Sergeant C. D. McCoy of Reger's Battalion wrote to his mother that "I am too much worn out to write—

only wanted to let you know that I am safe." He also informed her that "the slaughter on both sides was awful—dead men are lying all over the hill above us."

"We had a tremendous hard fight here on the 13th of the present month," Charles Delay of the 25th Virginia wrote to his parents, "but the Lord being our helper, we came off more than Conquorer." He added, "I received a musket ball in my overcoat, and a buck shot in my coat sleeve, but did not tough the hide anywhere, although I was in the hottest of the fight and had the pleasure of running and seeing the Yankees run, instead of them running me. I thank God that I was neither killed or wounded, and am still in the land and among the living."

The day was one long to be remembered, according to William Batts of the 12th Georgia, who noted that "many a gallant and brave man bit the dust on bowth sides." Batts, who was wounded in the battle, expressed his bitterness toward his foes, writing, "they marked me and many others with a scar that will be with us as long as life endures."[225]

The Yeagers returned to find their home riddled with bullets; one corner of the building had been badly damaged by a cannon ball. The home was also filled with wounded soldiers, many of whom died and were buried behind the house with their comrades who had fallen in the fight. Corporal John Henry Cammack and others of Company C, 31st Virginia, were forced to bury their own; the company lost five dead, seven wounded and one captured in the battle out of forty-two present. Casualties from the company were buried in a common grave. After the bodies were covered, Cammack commanded a party that fired a final salute to their fallen comrades. A coffin was prepared for Private John T. Wood of Reger's Battalion and his remains were interred on Sunday, December 15, with a service conducted by the Reverend William McIlvain, chaplain of the 44th Virginia. Captain Lilley sent Wood's letters and a lock of his hair to his family and commented that "though a stranger to most of us when he joined the company, he had since, by his true manliness of character and pleasant disposition, won the friendship of all."[226]

It took longer for the bodies of the Federal dead to be buried, since most were returned home to Indiana, Ohio, and Pennsylvania. On December 17, twenty-nine Confederate prisoners reached Philippi after a rugged march through ankle-deep mud. Described as "a hard looking set," they were given some meat and crackers despite the sentiment of their guard, who stated "they must suffer for their wickedness."

Two officers of the 9th Indiana were granted a fifteen-day leave of absence "for the purpose of conveying home the bodies of the soldiers of said Regiment." The remains of Josie Gordon of the 9th Indiana were interred in Indianapolis exactly one week after the battle. The coffin containing Gordon's body was draped with an American flag and a large procession of mourners trailed the hearse. A company of his father's regiment, the 19th United States Infantry, fired a farewell salute and the benediction was delivered by the Reverend A. L. Brooks.

Discovered in the young man's coat pocket after his death was an unfinished, blood-stained letter to his father, in which he wrote,

> *You seem to be at a loss, dear father, to understand my motive for volunteering, but I think, if you will remember the lessons which for years you have endeavored to impress upon my mind, that all will be explained. When you have endeavored ever since I was old enough to understand, you, to instruct me—not only by precept, but by example—that I was to prefer freedom to every thing else in the world, and that I should not hesitate to sacrifice any thing, even life itself, on the alter of my country when required, you surely should not be surprised that I should, in this hour of extreme peril to my country, offer her my feeble aid.*

General Robert H. Milroy wrote of Josie Gordon, "He died as only a brave soldier can meet death—in the front rank of the battle, and in the imminent deadly breech . . . Having been a member of my military family since the commencement of the present campaign, his many amiable qualities had endeared him to me, and I loved him as a son, and his death has created a vacuum in that family that cannot be filled."

Stephen S. Harding, soon to be appointed governor of the Utah Territory, also noted Gordon's ardent patriotism, stating, "How sacred is that offering. How it should remind the living of the worth of our liberties so baptized in the blood of one so young, so beautiful and so brave. The soil made red with so priceless a sacrifice shall bear, in time, naught but the fruits of human liberty; and the youths and maidens of a future generation shall point out the consecrated spot."[227]

Despite the small number of troops engaged, the Battle of Allegheny Mountain was one of the most closely contested struggles of 1861, and the bloodiest to date in western Virginia. According to Jacob Pinick of the 32nd Ohio, "It is said to have been the hardest fought battle that has yet been fought in Western Virginia. I know it was that hard that I don't care about being in another."

Casualties at the Battle of Allegheny Mountain were fairly equal, although those of the Confederates outnumbered Union casualties in large part because of the large number of the former captured during the battle. Documents reveal Union losses of thirty killed in action, thirteen who died from wounds sustained in battle, ninety-seven others wounded, and two listed as captured or missing in action, for a total of one hundred and forty-two.

Colonel Edward Johnson's Confederate force lost twenty killed, fourteen who later died from wounds, ninety-three others wounded, and thirty-four captured or missing, totaling one hundred and sixty-one casualties. They had succeeded in repulsing a determined attack but ironically their victory at Allegheny condemned them to a brutal winter of occupation on the summit.[228]

As news of the battle drifted to the North, newspapers proclaimed a great Union victory, with the Confederates reportedly in retreat toward Staunton. Gradually, however, more accurate reports began to appear and Federal supporters learned how costly the attack had been to no apparent avail. Several factors contributed to the failure of the Union expedition. First, the intelligence provided by Confederate deserters, although accurate as to the number of troops on Allegheny Mountain, was certainly incorrect in asserting the Confederates were prepared to surrender.

A second critical mistake was the decision to attack the fortified camp without artillery. The fact that seventy-five artillerymen marched to the summit without guns expecting to capture those of the enemy exhibits an obscene overconfidence in the ease with which Camp Allegheny could be taken.

Another factor that may have influenced the outcome of the battle was the lack of Union officers on the battlefield. It would have been more prudent to attack after some of the officers returned from leave, but Milroy apparently believed he could not wait. The scarcity of Union leadership on the field was critical. The resolve of the Federal attack seems remarkable considering the arduous and lengthy march and the soldiers' lack of sleep. As one Union soldier noted, "We never closed our eyes from Wednesday night until Saturday morning at 4 o'clock."[229]

The primary failure in the execution of the offensive was the inability of the two Union columns to coordinate their attacks on opposite ends of Camp Allegheny. The column that advanced on the Confederate right was forced into battle sooner than it wished due to the tenacity of some pickets from Hansbrough's Battalion, whose rapid return to camp caused the Union soldiers to reveal their presence.

Moody's column, which attacked the entrenched Confederates on the left, was slowed by fallen timber and the poor condition of the road, which caused them to fall behind schedule. By the time they reached the field of battle, the fighting on the right had nearly ceased and the Confederates were able to concentrate their entire force on the second column, having successfully repulsed the first. Milroy's plan was bold, and the Confederates would have been hard-pressed to hold their camp had the two columns attacked simultaneously, but the timing was exceedingly difficult to coordinate, particularly in such mountainous terrain.

Both newspapers in the North and Union soldiers blamed General Robert H. Milroy for the failed attack. The *Zanesville City Times* noted that Milroy "showed considerable trepidation." A member of the 13th Indiana, Sergeant Frank Ingersoll, was forced to admit that "we was whipped" and blamed Milroy for placing "to much confidence in them damned rebel deserters."

Upon reflection, Sergeant James M. King, also of the 13th Indiana, was angered by the poor leadership and lack of planning for the fight at Camp Allegheny. He wondered "what kind of Generalship is it, that takes a force of 1700 infantry, without artillery and scarcely ammunition for half a days fighting" to attack such a well-entrenched foe equipped with several batteries. King hoped "that this reverse may at least be the means of teaching our officers another lesson." It is ironic to note that had Milroy waited only a few days longer, his command would have taken Camp Allegheny by default.[230]

Some Northern newspapers reported that the fight was lost due to the cowardice of the 25th Ohio and the 2nd (West) Virginia. Although Milroy admitted that elements from these regiments, as well as others, were routed by a Confederate charge and did not return to the field of battle, he stated that no regiment should be singled out for criticism. While several letters written by Indiana soldiers also accused the troops of the 25th Ohio of fleeing the battlefield, a look at the casualties suffered by this regiment puts matters in a different light.

Out of four hundred soldiers, the 25th Ohio lost eleven killed in action, five who died from wounds, and fifty-two others wounded, almost as many casualties as all other Union regiments combined. Certainly this regiment should not have been accused of being responsible for losing the battle. As one of the soldiers noted the men of the 25th "fought like heroes but many a poor fellow of our Regiment bit the dust." Nevertheless, the cowardice of some troops combined with the continual trickle of others carrying wounded comrades to the rear weakened the Federal attack on the Confederate right.[231]

Many contemporary Union accounts blamed the attack's failure on a deserter from their own ranks. In his official report Milroy wrote that "we have since learned from prisoners that they were fully apprised not only of our coming, but of the plan of attack, two days before, by a deserter from our camp." While in many official reports officers blamed anyone but themselves for failure, in this case evidence appears to substantiate Milroy's claim. The accused was Bill McGee of the 25th Ohio, who reportedly deserted and went to the Confederate camp to warn of the attack. An Indiana soldier wrote that McGee was a scout sent out by Milroy to spy on the Confederates who instead told them all he knew. McGee was captured and imprisoned at Huttonsville. "I don't know what will be his doom," the Indianan concluded, but "he ought to be shot."[232]

While Union supporters sought a scapegoat, the Confederacy was delighted at the victory gained by the small army entrenched in the mountains. One reporter wrote that the gallantry of the Confederate defenders "rivaled the bravery of the Imperial Guard of Napoleon."

Ten days after the battle, Secretary of War Judah P. Benjamin wrote to Colonel Edward Johnson, informing him of President Davis's pleasure and gratitude for the victory. Davis had readily agreed with Benjamin's suggestion that Johnson be promoted to brigadier general. Johnson deserved the promotion, for he handled his troops admirably. By refusing

General Edward Johnson, who commanded the Confederate force at Allegheny Mountain during the winter of 1861-62. Library of Congress, Prints & Photographs Division, Civil War Photographs [LC-DIG-cwpb-07580]

to commit his entire force to the struggle on the right, Johnson retained the initiative, ordering troops into the fray as needed but remaining prepared to confront an attack from the opposite side of the camp. As a result of his leadership and the defensive works on Allegheny Mountain, a Union victory would not necessarily have been assured even if their columns had attacked simultaneously.[233]

Johnson's promotion produced scorn from one soldier and a frantic quest to influence who would replace him as regimental commander. One member of the newly promoted general's former regiment, the 12th Georgia, commented sarcastically, "He will do for the world to honor having all the savage ferocity of one born to spread devastation around."

Several of the company commanders signed a letter sent to Confederate Vice President Alexander Stephens requesting that Lieutenant Colonel Z. T. Conner not be promoted to colonel of the regiment and seeking to have Captain James H. Deshler appointed to the command. No reason was given for the dissatisfaction with Conner, but the captains threatened to resign and claimed if he was ascended in rank the regiment "will be completely demoralized." Conner was disheartened by this insubordination, but was eventually appointed to head the 12th Georgia. Also unsuccessful was a later plea to have the regiment withdrawn from Allegheny Summit.[234]

Johnson issued a congratulatory address to his command on December 16 praising them for their victory of three days prior:

> *Georgians and Virginians, you have met the same enemy you met at Greenbrier river on the 3d of October, and with an equally glorious result. Whilst we have abundant cause to thank God for this victory, let us not forget the gallant dead who fell by our sides, and whom we buried on Alleghany. Remember their gallantry, and emulate their example.*

One Confederate wrote that a victory attained at so high a price could not be celebrated: "I cannot exult over our victory. Such work is a shock to human nature. We lost too many brave men to rejoice over our victory."[235]

The Confederate force on Allegheny Mountain, which had been prepared to withdraw eastward, now appeared to hold an important position and was ordered to temporarily remain in place. On the day following the battle, Sergeant C. D. McCoy noted that the Confederates had received marching orders to Staunton as soon as transportation could be procured for "'field service in another line.' Yesterday's proceedings will delay our movements and may prevent our going entirely—Don't know where we are going."[236]

Immediately after the battle concluded, Johnson wrote to C. L. Stevenson, Loring's adjutant, "I trust immediate action will be taken relative to this position. Under recent orders, I have sent to the rear a large quantity of ordnance and ordnance stores. I have all along contended that this place would be occupied if we abandoned it." Johnson also informed Stevenson that he was preparing for another attack on the camp, although he believed this unlikely, and sought a decision so that he could either re-supply his force or continue to send stores to the east in anticipation of a withdrawal.

An answer was not immediately forthcoming and Johnson wrote again on December 15, seeking a decision:

> *Immediately after the fight I ordered the transportation of stores from this place to stop, and no more trains to be sent to this place until further orders. I am strengthening my works, and I trust that something decisive will be determined upon, so that I may know what to do.*[237]

Two days after the battle, Adam Stephenson, an attorney and apparently influential citizen of Monterey, wrote to J. W. Brockenbrough concerning the planned withdrawal from Allegheny. Stephenson and the rest of Monterey's populace feared that, without a Confederate force guarding the summit of the mountain, they might be "subjected to the ravages of the notorious Yankees."

Brockenbrough responded to Stephenson's request for the continued presence of a Confederate force on Allegheny Summit on December 18, when he wrote to Secretary of War Judah P. Benjamin. He stated that, had Johnson's army been defeated in the recent battle,

> *there would have been no obstacle to an onward march to Staunton, in the very heart of the Valley of Virginia. I trust there is no foundation for the rumor which has so much agitated the population of that part of the valley, that it is the purpose of the Government to deprive them of all protection.*[238]

Brockenbrough's plea was unnecessary, as General Loring had finally made a decision concerning the Confederate force guarding the turnpike over Allegheny Mountain. On December 17, Loring sent a dispatch to Confederate Army Adjutant and Inspector General Samuel Cooper, writing that, "in consequence of the necessity of meeting the enemy at Alleghany, and the uncertainty of their movement, I have determined to keep the command of Colonel Johnson where it is for the present." Loring believed that the determination of the enemy to occupy Camp Allegheny revealed their intention to move a force across the Alleghenies to link up with General Benjamin F. Kelley's force via Moorefield.[239]

On the same day Loring wrote to Colonel Edward Johnson, formally revoking the order to withdraw and informing him that his force would remain on Allegheny Summit for the foreseeable future. Loring ordered supplies sent to Johnson's command, which in addition to the supplies on Allegheny would be sufficient for two months. The adjutant for the commanding general of the Army of the Northwest wrote,

> *It is the intention to hold Allegheny Pass and the country in its vicinity, and you will please dispose of your command with a view thereto. Circumstances, however, may render it advisable to detach it for service with the forces now moving toward Winchester, and the general desires that you will keep it constantly and fully prepared for such an emergency.*

As Major John D. Ross noted succinctly, "We were ordered East, but the battle changed the programme."[240]

Following the Battle of Allegheny Mountain, there was little military activity in the region during the winter of 1861-62. General Thomas J. Jackson's offensive against the Federal garrison at Romney was launched in January 1862. Minus Johnson's brigade, the rest of Loring's Army of the Northwest joined Jackson at Winchester and on the first day of 1862 the combined force moved into northwestern Virginia. The Confederates reached Romney on January 14 but found the town deserted. Jackson's attempt to capture the Federal garrison failed, but he was determined to hold Romney through the winter months, threatening the B & O Railroad. Consequently, Loring's command was ordered to remain in Romney while the Stonewall Brigade was withdrawn to Winchester. The resultant protests

Wartime sketch of Romney, winter of 1861-62, by Alfred R. Waud. Library of Congress, Prints & Photographs Division, Civil War Photographs [LC-DIG-ppmsca-22434.]

by the officers of Loring's army, however, led Confederate Secretary of War Judah P. Benjamin to order Jackson to withdraw the Army of the Northwest from Romney, an intervention that prompted the general to submit his resignation in protest.

The Union attack on Camp Allegheny failed to dislodge the Confederates, but it did have a direct effect on Confederate strategy and troop movements in the region. The assault so alarmed General W. W. Loring, who feared that Federal forces would occupy the pass through the Alleghenies in a first step toward an advance on Staunton, that he countermanded his earlier order withdrawing Colonel Edward Johnson's command eastward toward Staunton and ordered them to remain in place for the foreseeable future. Loring's orders were directed by his belief that if the pass over the Alleghenies was left unguarded, a Federal advance would occupy Camp Allegheny and possibly advance eastward along the Staunton-Parkersburg Turnpike toward Staunton. Although the absence of Johnson's brigade appears to have had little effect on Jackson's offensive, once again, the actions of Confederate military leaders had been guided by their perception of Federal designs in the region.

CHAPTER SEVEN
WINTER WORK

On December 21, four or five Union soldiers were reported to be eating about four miles from Camp Allegheny. One company from the 12th Georgia and another from the 52nd Virginia were sent to capture them. Swift marching brought the Confederates within sight of the enemy on the far side of the Greenbrier River, but according to Major John D. Ross, "they took to the mountains and got away." Their guide, John Slayton, "who has led them on all their scouting and marauding expeditions, was with them, and it was for this reason we were so anxious to get them. He is a Union man and lives about four miles from here."

Another alarm was raised in the camp two days later. The brigade was ordered to the entrenchments where they remained for four hours awaiting an attack that never came. It was later determined that the alarm was caused by some of the Confederate soldiers slipping past the pickets to search for liquor. When they returned to camp, they were discovered and fired upon by the sentinels. Such alarms occurred with frequency during the winter of 1861-62, producing no large-scale fighting but plenty of disgruntled soldiers. [241]

Much of the military activity during the next few months took place along the Dry Fork of Cheat, a wilderness inhabited by few humans. "This country, though of no value to us for any purpose, is, in my judgement, of great value to us in a military point of view," one Union officer wrote of the Seneca/Dry Fork region. Noting that cavalry could travel from Corrick's Ford, in Tucker County, to Seneca, in Pendleton County, approximately 31 miles in eight hours and the infantry in twelve, the officer continued,

> *thus by starting at 6 in the evening at Seneca troops can be run in within 12 miles of the pike, can cut off all possible means of the report of their advance reaching any point on the pike for one day and then the next night can make a run onto the small force that is left to hold the important points on the road and can either hold them until a heavy force can reach them or they can commit*

> *heavy depredations and then return to the mountains before a force sufficient to take them can be brought from any point on the pike and this they contemplate as soon as the weather will permit. The number of men on this river from the mouth of Black Fork to Soldier White's is 45 and about 23 of them are loyal and about 10 more would be if they were not afraid of the rebels, the rest are the perfect outlet devils in the name of home guards.*[242]

Although control of the region was in fact insignificant, the small-scale struggle conducted here was bitterly fought. Fighting had occurred along the Dry Fork in the fall of 1861, but the intensity rose following the Battle of Allegheny Mountain. On December 19, Captain George Latham of Company B, 2nd (West) Virginia Infantry, stationed at Belington, in Barbour County, ordered thirty-eight men under the command of 1st Lieutenants Fabricius A. Cather and Daniel Wilson to undertake an expedition to the Dry Fork, where a Home Guard company headed by Captain Samson Elza was purportedly gathered.

After a rapid march covering sixty-five miles, the Union troops reached the Confederate camp, which they surrounded on the morning of December 22. At daybreak they stormed forward only to discover that the enemy had dispersed, leaving only some papers and a number of musical instruments, including a fife, two fiddles and a dulcimer. As the sun's rise birthed the first light of day, a rapid fire was opened on the Union detachment by the Home Guard company known as the Mountain Tigers, who were hidden behind rocks and trees along the mountainside. The Federals returned the fire, but the local men were careful to "keep either out of sight or out of range of our 'off hand cannon (rifled muskets)'." The steady fire did encourage the Mountain Tigers to gradually disperse.

Company B suffered no casualties; that of the Home Guards is uncertain, but one man was reportedly seriously wounded during the brief and otherwise harmless encounter. The Union party then marched into Pendleton County, twenty-five miles from Camp Allegheny. After disarming some Confederate sympathizers and taking the horse of the most notorious, the detachment returned to Belington on Christmas Eve, undoubtedly exhausted from both the physical exertion and the mental strain of marching through enemy territory.

Union Colonel George Latham of the 2nd [West] Virginia Infantry. Library of Congress, Prints & Photographs Division, Civil War Photographs [LC-USZ62-100466]

Meanwhile, fearing the worst for the men of his company, Captain Latham obtained permission for a relief expedition into the region. Thirty-six men of Company F, 3rd (West) Virginia Infantry under 2nd Lieutenant Asa Coplin left Belington and took a shortcut through the mountain wilderness. Within eight miles of the Dry Fork of Cheat, Coplin received word that a Confederate force was blocking the road. Without hesitation Coplin led his men back to Belington, to Latham's great consternation. Captain Latham gathered eight men of his company and on the afternoon of December 22 departed for the Dry Fork. The small force moved at a forced march for a day and a half, reaching the scene of the skirmish in the early morning hours of December 24. They found the rebel quarters deserted and learned that the expeditionary force was returning to camp. Latham's small party remained in the neighborhood for a day and escaped their dangerous position by marching back to Belington, which they reached on December 25, having celebrated Christmas along the march.

One participant stated that the expedition "struck a perfect panic into the rebels of that region, who had considered the mountains all their own. We traveled fifteen miles above where any Union soldiers had ever been before, and there were two hundred in the party that even went within that distance of the scene of our operations. Long will Company B of the 2nd Virginia Regiment be remembered in these hills and mountain fastnesses." Although the expedition was undeniably daring, Home Guard activity remained a problem for the Union troops in the Cheat Mountain region.[243]

At the end of December, General Milroy ordered a more ambitious undertaking. At 1 p.m. on New Year's Eve, Major George Webster led a detachment of four hundred troops of the 25th Ohio Infantry from Huttonsville on an expedition to Huntersville, a major supply depot for the Confederate Army of the Northwest. Eight wagons and two ambulances accompanied the column, the former carrying forage and six days' rations for the men.

The guide for the expedition was a Dr. Samuel Gibson, a local Union man, who was vilified by Confederate supporters and military leaders in the region. Gibson had been taken prisoner by the Confederates but escaped and conveniently showed up in the Union camp a few days before the raid commenced. After a march of two miles, darkness set in and the Union soldiers made camp in an open field four miles north of Camp Elkwater.

There had been no room in the wagons for their tents so the soldiers slept around campfires under the stars. They were awakened at 5 a.m. by the call of a bugle and enjoyed a hearty meal of coffee, salt meat, and hard crackers before continuing their journey.[244]

Marching south on the first day of 1862, Webster's command was joined at Camp Elkwater by three hundred troops of the 2nd (West) Virginia Infantry. The expeditionary force encamped for the night at Big Spring, where they found pots, kettles, tents, gunlocks, and other various camp equipage which had been left behind by the Confederates when they withdrew in late September.

On the following day the Union force continued the advance, joined by thirty-eight troopers of Bracken's Cavalry. Fourteen miles from Huntersville, at the base of Elk Mountain, the column reached a blockade of felled trees which stretched for nearly three-quarters of a mile, constructed by the Confederates to prevent wagon travel on the road between Huttonsville and Huntersville. Webster's command stopped for the night, but early on the morning of January 3, the supply train was left with a guard of fifty to one hundred of his most disabled men, while the main body of troops was guided around the blockade by making a wide detour to the east.[245]

"Everybody was astir at Huntersville on the morning of January 3," recalled one Confederate, and the men were undoubtedly a bit edgy. Wagon-master Robert Marshall and Lieutenant J. N. Potts, the post commissary, went to the quarters of Captain D. W. Alexander and urged him to order the immediate removal of the thousands of dollars of provisions and stores on hand.

Alexander "utterly and stubbornly refused, saying that he would not, under any circumstances, make preparations to retreat." Alexander also failed to make proper preparations for the expected arrival of the Union force, ordering no inspection of his troops. There were 250 Confederates at the Huntersville post, but with a number absent on furlough, and others sick or drunk, only 150 men would be available to defend the town.[246]

Proceeding as quietly and cautiously as possible, the Union force pushed on toward Marlin's Bottom (present-day Marlinton; also referred to during the war as Greenbrier Bridge), where two Confederate cavalry companies and a militia force were gathered, a total of about 250 defenders. Companies E and G of the 25th Ohio, serving as the advance guard, quickened their pace to double-quick when they spotted about thirty Confederate pickets guarding the bridge over the Greenbrier River.

At the same time, Webster ordered Lieutenant Hugh Delzell of Bracken's Cavalry to ford the river with his troopers and gain possession of the road beyond the bridge, trapping the small enemy force. The Union cavalry made a "furious and dashing charge," which was witnessed by the Ohio soldiers. One wrote, "From our position, three quarters of a mile distant, we could see the secesh darting out from the bridge like rockets, and galloping toward a little run, up which led the road to Huntersville. The valley was a meadow, without fences or any other obstacle, and the point to be gained was the entrance to the ravine." Some of the Confederate horsemen reached the gully, but most were forced down the valley toward the Gauley road, leaving provisions and forage in their wake.[247]

Lieutenant Potts and the other Confederates were awaiting news of the Federal advance, when suddenly Captain Jacob W. Marshall, "with a handful of brave men from the neighborhood of Hillsboro, rode into town, and reported that they had encountered a force of the enemy" at Greenbrier Bridge. Marshall estimated the Union force to be about 1,000 and noted, "they were moving upon us, and must be by this time within two miles of town."

Company F of the 31st Virginia Infantry, known as the Randolph Riflemen, was quartered in Huntersville but did not have a commissioned officer with them. Captain Alexander ordered Lieutenant Potts to take charge of this unit. Potts had originally been a member of Company F but had been detailed a few weeks earlier to act as commissary of the post. When he expressed concern about oversight of the commissary, Alexander responded, "You take care of the company and let the stores take care of themselves." Potts formed the Virginia company and made "a hasty inspection" of their arms. Alexander sent his company of Tennessee cavalry, under the command of Lieutenant Knight, to meet the Union advance and ordered Potts to move his company a few hundred yards below town and await orders from Knight.

Captain Archibald Richards of the Bath Cavalry formed his company and was ordered to move forward and report to Knight. The Pocahontas Rescuers, a company of infantry camped nearby, was apparently forgotten until Potts convinced Alexander to send for them.[248]

Company F of the 31st Virginia moved at double-quick to the high ground at the eastern end of William McLaughlin's farm. From this position, the Confederates could see the enemy in line of battle. Just below McLaughlin's house, Lieutenant Knight turned to his Tennessee cavalrymen and asked, "Now who will dismount here and fight for his country?"

Captain Richards, whose Bath Cavalry had arrived on the scene, "told him (with an oath) I was one that wouldn't." Despite Knight's pleas for his men to take positions behind rocks and trees just below McLaughlin's, the cavalry quickly withdrew to Huntersville, forcing Potts and his Randolph Riflemen to fall back. Captain Alexander then ordered the evacuation of the town, and his command moved up Knapp's Creek and fell back to Monterey.[249]

Union Major George Webster moved rapidly toward Huntersville, six miles distant, leaving Captain Jeremiah Williams with a detachment of fifty men to guard the bridge. Williams had the ends of the bridge blocked with heavy logs and "pierced loopholes at each post, determined to hurt somebody before leaving our position." A Confederate force was seen nearby, but when Williams ordered six of his men to advance on them, they withdrew.

Meanwhile, as Webster's men marched along a flat section of ground two miles east of the bridge, Confederates posted behind a ridge to their front fired on the advance companies, seriously wounding Private Oliver P. Hershee of Company E, 25th Ohio, in the arm. Organized resistance to the advance, however, did not occur until less than one mile from Huntersville, where Webster spotted a body of Confederate cavalry "at the extreme of a level bottom field, dismounted and posted over the brow of a hilly spur which jutted out into the field from their right, with Nap's (Knapp's) Creek on their left." Webster immediately deployed part of his regiment up the hill to their left to flank the cavalry, while the rest advanced straight ahead.

The Confederate cavalry opened fire on the approaching Federals who returned the greeting. Discovering the body of Union troops threatening their right flank, the troopers withdrew to their horses, mounted, and retreated. Webster's command followed, crossing Knapp's Creek and moving to within half a mile of Huntersville, where they spotted the Confederate cavalry posted in a bottom, protected by some old log buildings. Companies A and B of the 25th Ohio were immediately deployed into line on the right at the base of the hill while the 2nd (West) Virginia flanked the enemy's right and attempted to cut them off from the town. This movement was answered by the Confederate cavalry, which began to fire upon the Federals. The two companies of the 25th Ohio responded and the Confederates soon retreated into Huntersville, narrowly escaping the Union snare.[250]

After resting for a moment, Major Webster formed his men into line of battle and ordered them to fix bayonets. Waving his sword, he exclaimed, "One gallant charge, boys, and Huntersville is ours!" The Union soldiers advanced double-quick up an ascent "with a shout that made the welkin ring." "Our shouts that rent the air seemed to confound and dismay them," one Federal wrote, "for they seemed not to hesitate, but their flight was instantaneous and precipitate, our gallant Major was the first to ascend the eminence on which the town is situated, and drawing his sword from its scabbard, he waved it defiantly over his head, which occasioned another shout that made the mountains echo round, and increasing our speed, we were soon traversing through the deserted streets."

Lieutenant Charles B. Jones of the 25th Ohio was the first Union soldier to reach the center of town. He hauled down a "secession flag" and the "Stars and Stripes were taken up to the belfry of a church which was in the town, and unfurled to the breeze, at the sight of which the loud huzzas that rent the air were really astounding." The retreating Confederates fired ineffectually before departing from Huntersville, leaving the Federals in possession of the town.[251]

Huntersville was deserted; houses were broken open, and goods scattered along the road. Large quantities of supplies were found, including 347 barrels of flour; 300 gallons of vinegar; 43 boxes of soap (545 pounds per box); 30,000 pounds of salt; 15,000 pounds of bacon; 320 pounds of candles; 19,800 pounds of coffee; 153,000 pounds of beef; and a large quantity of rice. Webster estimated that the supplies were worth $30,000.

Unable to transport the goods back to camp and fearful of a counter-attack, at about 3 p.m. Webster ordered the large brick commissary building in which the supplies were stored to be razed. Also burned were other buildings filled with oats, corn, and more clothing. The Federals seized a large number of Sharp's carbines, sabers, horse-pistols, and some army clothing, some of which were carried back to camp, while most were destroyed.

They also discovered about thirty barrels of rye whiskey. A member of the 25th Ohio wrote: "The boys took a snifter for the stomach's sake and committed the balance to the streets, so that the secesh, if they should return to see the evil of their ways, could have whiskey and water for their infirmities."[252]

After remaining in Huntersville for about two hours, the Union force began the long march back to their camps in a mixture of rain and sleet described as "drenching." Before departing, Webster nailed the Stars and Stripes to the top of the courthouse. Three hundred additional soldiers from the two regiments met Webster's command at Big Spring, where they had been dispatched to guard against any attempt by the Confederates to cut off the Union force, but the Federals completed and returned from their mission before a response could be initiated.[253]

While the Union advance on Huntersville was underway, on the evening of January 2, the Confederates at Camp Allegheny were stunned to see fires burning at their former camp at the base of the mountain. One Confederate soldier claimed the fires extended at least a mile. Brigadier General Edward Johnson ordered his brigade into the trenches and immediately dispatched a message to Lieutenant Colonel Michael G. Harman in Staunton, informing him that a large Union force was at Camp Bartow and that he expected to be attacked on the following morning. Johnson expressed confidence that his force could hold, but he feared the enemy would circle behind him via Huntersville. In fact, earlier in the day he had received a dispatch from the commanding officer at this vital supply post, who expected an immediate attack upon his small force of about 250 soldiers.

Johnson ordered the 44th and 58th Virginia Regiments to advance toward Camp Allegheny and requested additional reinforcements. The 58th, which had been encamped at Forks of Waters, was ordered up to Monterey while the 44th moved forward to Hightown in Crab Bottom, about midway between Monterey and the Summit. These regiments were weak, however, with a combined strength of only six or seven hundred. After marching all night both arrived at Camp Allegheny on Friday morning, January 3, at 8 a.m.[254]

Shortly after ten o'clock on January 3, Johnson sent the 31st Virginia and 12th Georgia to Camp Bartow to learn the enemy's strength. To their surprise, the former Confederate camp at the base of Allegheny Summit was deserted. Fortifications had been destroyed and buildings burned, but the Union force which had rendered the destruction had departed, leading some to believe that the raid on Bartow was in fact a feint, covering a larger movement elsewhere. The return march of these regiments was treacherous, as a freezing rain made the road "icy slick." Many of the troops were injured from falls while others crawled their way to the top of the mountain. The 44th and 58th were ordered back to Hightown and Monterey respectively, both regiments reaching their destinations on Saturday evening after a difficult march "over a muddy and detestable road." Their arrival eased the fears of those stationed in the latter town, who expected the next Union blow to fall upon them.[255]

News of the raid on Huntersville disturbed the Confederate commands at Monterey and Camp Allegheny. Johnson sent out scouts along all roads in the surrounding area to prevent a second attack on his camp. Some of the Confederate soldiers expected to withdraw to the east following news of the Union triumph.

William W. Hardwicke, 2nd lieutenant of the Lynchburg Artillery, feared that Johnson's "pygmy army" would be forced to march to Staunton to prevent being cut off by the Union force, which the Confederates believed numbered in the thousands or even tens of thousands. He commented that the Confederate government "has treated us shamefully, leaving us in the bleak mountains with a small force to face the enemy of 20,000, who can surround us at any time and cut off our supplies." This same soldier expressed confidence in the defensive strength of Camp Allegheny, stating that if the Union force failed to cut off the supply line and "will attack us at this point, we will give them a cordial reception."[256]

The Confederate cavalry companies that had been driven from Huntersville arrived in Monterey on the evening of January 4, while the infantry companies were strung out along the road to the town. Not only had these troops lost their tents and baggage, they were also hungry, and they were angered when they remembered that just prior to the raid they had finished salting about 125 head of cattle.

A strong picket was sent out from Monterey to travel sixteen miles in the direction of Huntersville and preparations were made to repel an attack. In addition to the 58th Virginia, three militia and two cavalry companies were present in the town, forming a force of over one thousand.

One Confederate soldier expected simultaneous attacks on Monterey and Allegheny Summit; he also expressed confidence in the strength of the fortifications at the latter camp: "I felt and feel very little apprehension as to the result of an attempt on Alleghany, even by an overwhelming force, as our fortifications are thoroughly completed there, and the men are anxious and eager for an opportunity to use them." Efforts to improve their winter quarters also continued. Details were sent from camp each day to make clapboards for cabins and a stable large enough to hold 150 horses.[257]

Johnson had about twelve hundred effectives on the summit. He stated that if the Union force advanced toward Monterey, "re-enforcements are imperatively required." On January 4 Lieutenant Colonel Harman telegraphed Adjutant General Samuel Cooper that Johnson needed two regiments to guard against being cut off from his supply line. Cooper apparently passed the request on to General W. W. Loring, who was at Winchester. Loring wrote to Cooper agreeing that Johnson needed aid but declaring, "he will have to be re-enforced from elsewhere." There would be no reinforcements, but the threat subsided and the stalemate resumed.[258]

The Union soldiers of the 25th Ohio and the 2nd (West) Virginia Infantry regiments had undoubtedly read the accusations charging them with cowardice in the Battle of Allegheny Mountain. The daring raid on Huntersville not only redeemed their reputations, it also caused great alarm for the Confederate force still occupying the summit. General William S. Rosecrans expressed his thanks to Milroy and his troops but warned him to "keep on the alert, and not have the compliment returned."[259]

At 8 p.m. on the evening of January 3, three hundred of the 32nd Ohio under the command of Captain Jackson Lucy, who was known for providing medicine and medical attention to his company, had departed for Tucker County, where they were to root out "some horse thieves who had a fashion of coming down into the valley and stealing horses." One of the men wrote, "For miles of our way there was no road, the one wagon of the expedition having to be drawn up the precipitous mountain sides by the troops and again let down by ropes in the hands of the soldiers. The people of that wild, mountainous region had no use for wagon roads, as they had no wagons."

On Sunday evening, January 5, the Union party was fired on by about ten secessionists who were hidden along the side of a mountain on Shaver's Fork near the home of John Taylor. One member of the 32nd Ohio was shot in the finger, but he was the only Union casualty. Four Confederates were reported killed and several more wounded, but this account is most likely exaggerated. The Ohio troops captured three enemy soldiers, one a lieutenant, and a small commissary before returning to Cheat Summit. One of the Ohioans wrote, "The Secesh here won't fight us fair. They do all their fighting in the woods, but we are about as good as them at bushwhacking."[260]

Simultaneously, a smaller Federal expedition was ordered into the Dry Fork region of Randolph County in search of Bill Harper and his gang of bushwhackers, the Dixie Boys, who had been stealing cattle to supply Confederate troops. Captain George Latham of Company B, 2nd (West) Virginia, along with nearly twenty of his men, was joined by Lieutenant A. J. Weaver and a detachment from Company K.

On January 7 they reached the Dry Fork of Cheat. Weaver and his detachment stopped at John Snyder's farm house, but Latham's men continued up the valley about two miles. After spending the night at Armentrout's, they headed up the river nearly a mile to Harper's home, but found it unoccupied. The Union soldiers started back, but in an open bottom near Armentrout's residence, near the confluence of Gandy Creek and the Dry Fork of Cheat River, they were ambushed by the Dixie Boys, who were secreted in heavy timber. Latham's men heard the click of the enemy's guns and dropped to the ground to avoid the volley. The Federals then fell back to the cover of the river bank and began fighting back, "standing and crouching in the water through a thin ice."

According to a letter to the *Wheeling Daily Intelligencer*, the encounter was brief but the fighting fierce: "Though the number engaged were small the firing was so rapid that it was distinctly heard for eight miles. The parties were within thirty steps of each other when the fight commenced, and the rebels, owing to the superiority of their numbers and position, were so confident of success that they fought for a time like tigers, but were finally driven entirely off the field." Confederate Captain C. G. Barksdale, however, reported "the yankeys retreated back from the ground."[261]

The fight continued for about an hour, and according to a Union report, the casualties were heavy. The small Federal force had six men wounded, including one whose wounds were believed mortal. This loss of nearly half the Union party indicates the ferocity of the brief skirmish. Federal claims that six rebels were killed and several others wounded were exaggerated;

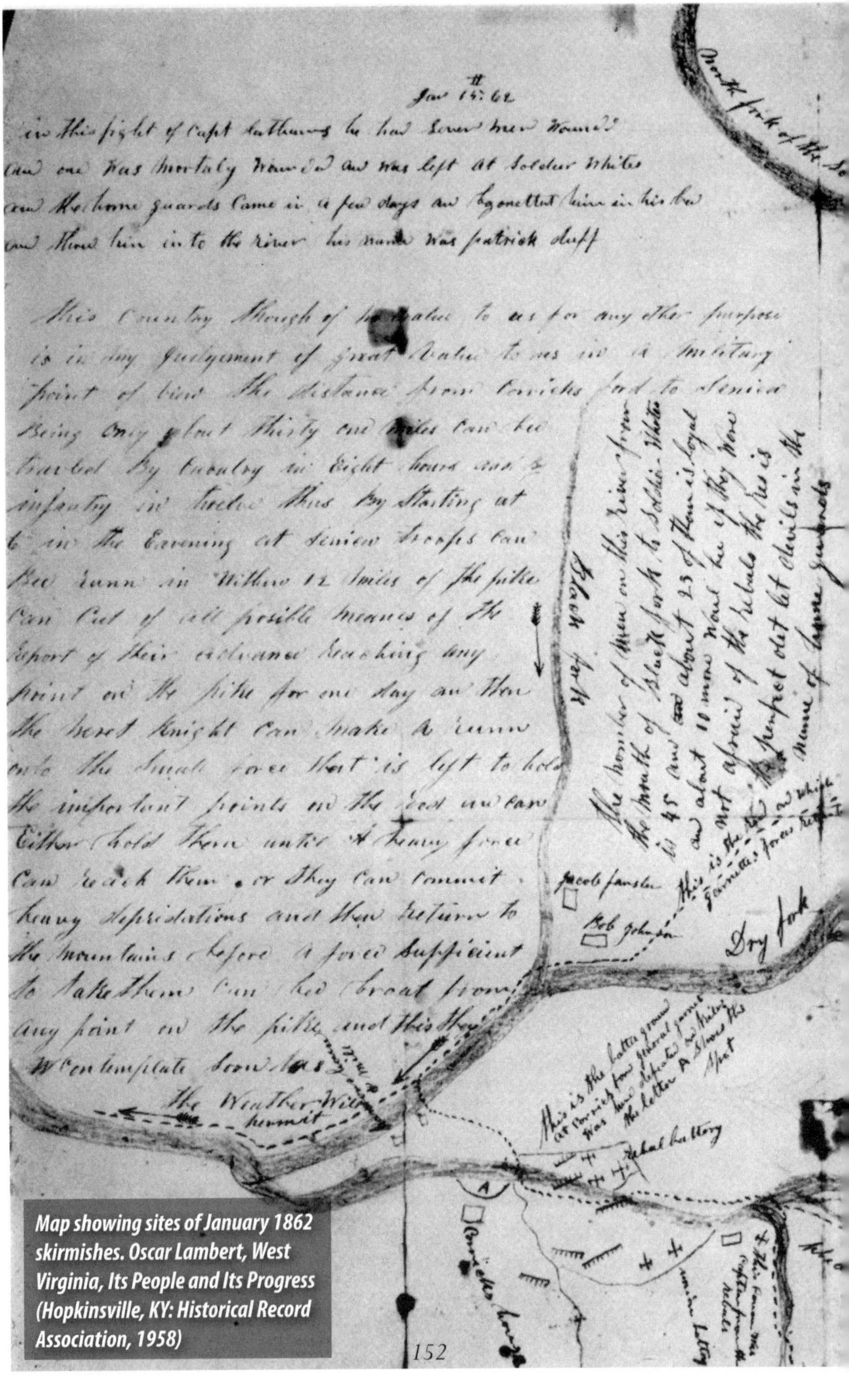

Map showing sites of January 1862 skirmishes. Oscar Lambert, West Virginia, Its People and Its Progress (Hopkinsville, KY: Historical Record Association, 1958)

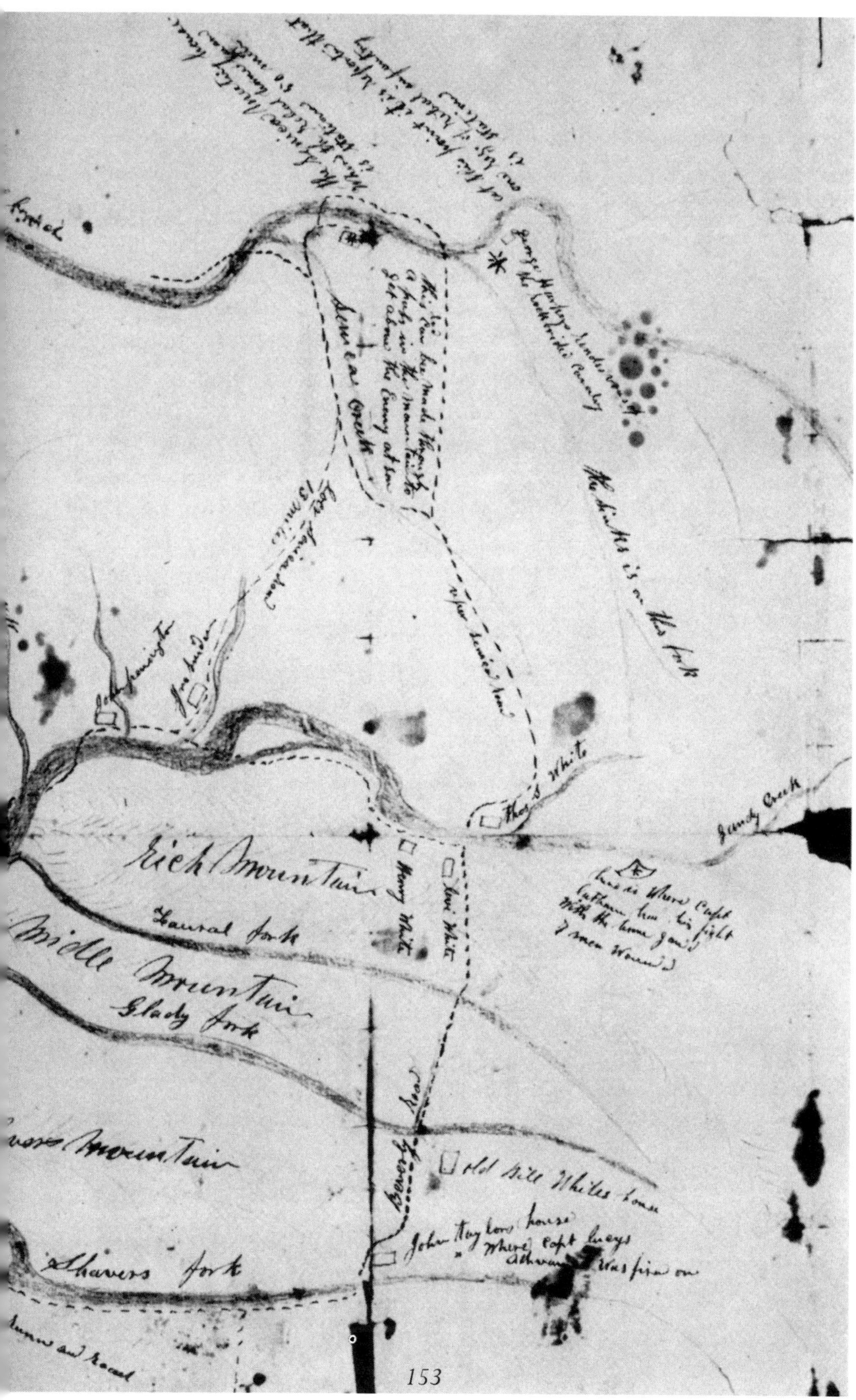
Rich Mountain
Laurel fork
Glady fork
Shavers fork
Seneca Creek
13 miles
Sandy Creek
Henry White
old Bill Whites house
John Taylors house

actually two of the Dixie Boys were wounded and one killed. Private Frederick Dopp, the oldest man in Latham's company, was shot through the body just above the heart and was unable to make the return journey. Dopp was taken to the Snyder home, where he was left in the family's care.

The Federal troops then seized horses from local residents and started down the river with the wounded men. The snow and rain beat down upon them as they rode through streams filled with ice and roads thick with mud. On January 11, they reached Rowlesburg, Preston County, and took the B & O Railroad to Grafton, where the wounded were hospitalized. After a brief rest, Latham and the remainder of his detachment rejoined their comrades on Cheat Summit.[262]

Confederate Captain C. G. Barksdale accompanied Captain Joseph H. Lantz's company into the Dry Fork region on the evening of January 8, but less than a mile down the road they met the Dixie Boys. Barksdale argued that both companies should return to destroy the Union detachment, but Lantz instead sent two men from each company to ride to Camp Allegheny on the following evening.

The roads in the region remained in nearly impassable condition, particularly from Webster to Beverly. In mid-January a member of the 32nd Ohio, stationed at Beverly, wrote that the "mud is about 6 inches deep in the streets and about as thick as buckwheat batter."

In lieu of military activity, the Union soldiers resorted to various ways of passing the time, including the placing of bets on how soon the war would end. One wrote that he thought "the war will be settled in about 3 months if England don't put her finger in the pot but I think she will not meddle herself at this time. If she will only wait till the family muss is settled, then I for one will volunteer to whip her but I hope it will not be necessary."[263]

Another Federal scouting party was sent to the Dry Fork region on January 19 to protect Union supporters who were attempting to organize and arm a local company. They stopped at a house to get something to eat, but after dining on the meal prepared by their hostess, they became violently ill. Suspecting the woman had poisoned their coffee, officers arrested her. This detachment brought in thirty head of cattle and several prisoners.

On the following day, General Milroy informed his wife that several Union citizens who had been imprisoned at Staunton had recently escaped and made their way to his camp at Huttonsville. From these men Milroy learned that:

> *the gathering of rebel forces at Monteray, (which I believe I mentioned in my last letter) instead of being for the purpose of attacking us was for the purpose of resisting an apprehended attack by us. That we gave them such a terrible handling on the 13th ult. that they have been in constant apprehension of our coming again and when they heard of our taking Huntersville they supposed our forces were coming right on to Stanton and the wealthy Secession citizens commenced moving out rapidly—they feel certain that I am coming and are everywhere along the line making preparations either to fight or run. Mostly the latter.*

The general also expressed his hope of an immediate advance eastward, writing, "If they will not let me leave this miserable region I think I will have the 9th [Indiana] brought back here with two or three other [regiments] and go on through to Staunton. I suppose our General will let me go out that way."[264]

Meanwhile, Confederate officials sought various means to maintain the health of the troops at Camp Allegheny. Medical officers of the 52nd Virginia sought the distribution of molasses to the men "constantly and liberally" for both medical and "palatable" reasons. It was thought not only to be important to the health of the men, but also "of great & peculiar value when taken in connection with the rice ration which the men cannot eat without some means of making it palatable." The soldiers ate fairly well during the winter season, having plentiful supplies of beef, pork, rice, sugar, coffee, and flour. A favored delicacy was beef brains, described by one Confederate as "a very nice dish like oysters."[265]

Despite their best efforts, the Confederate leadership was unable to prevent the inevitable scourge of disease. The 12th Georgia provides a case in point. In the battles of Cheat Mountain, Greenbrier River, and Allegheny Mountain, ten men from the regiment had been killed, thirty-four wounded and three captured. During their stay in western Virginia, however, 129 soldiers died from illness. A similar number were discharged from the regiment. This unit would participate in the war's bloodiest battles, including McDowell, the Richmond campaign, Second

Manassas, Fredericksburg, Chancellorsville, Gettysburg, the Wilderness, and Spotsylvania, but more lives were lost from disease while in western Virginia than in all of these battles combined.[266]

The leaders of the 12th Georgia blamed some of the health issues on their surgeon, Dr. Henry K. Green. On January 6, 1862, captains of nine of the companies of the regiment signed a complaint against Green, alleging that he neglected to attend to sick soldiers, failed to prescribe medicine and refused to fill his own prescriptions. Lieutenant Colonel Z. T. Conner, who had been given command of the regiment following the promotion of Edward Johnson, endorsed their accusations, stating that Dr. Green "has grossly neglected the men of the Command in many respects, due to his gross want of method and memory."

Secretary of War Judah P. Benjamin recommended that if the doctor had not been confirmed, he should be dismissed from the service. Green had been confirmed, however, and Benjamin therefore believed a court martial was in order. Dr. Green was notified of the charges on February 4 and a day later was defended by General Edward Johnson in a scathing dispatch. Johnson was infuriated by the charges against Green, whom he described as an "active and skillful surgeon," and thought the disgruntlement stemmed from Green's position as head of a board that examined applicants for discharge. Less than one in twenty were given a medical discharge, and Johnson believed Green's actions "may account somewhat for his unpopularity & the baseless charges that have been brought against him." He also noted, "No officer, medical or other, discharged his duties more fearlessly or faithfully than Dr. H. K. Green, who has been dropped without a hearing." Johnson's reply eventually got a response assuring the general that "no order has been issued by Adjutant General's Department droping his name from the rolls." A court martial was ordered, but no records were found. Green was on the roll of the 12th Georgia on January 30, 1865.[267]

Less than a week after the Huntersville raid, Confederate scouts were sent out from Monterey to Huntersville to determine if the Union force was still in the area. Captain D. W. Alexander's favorite scout, an old man named Wood, went near the blockade and learned from locals that the enemy was not in the area and had in all likelihood returned to Crouch's, just south of Huttonsville on the Huntersville-Huttonsville Turnpike.

Alexander requested that supplies for his troops be sent to Huntersville in anticipation of the return of a Confederate presence there. In a January 15 letter to General Edward Johnson, Captain Alexander passed on accurate intelligence on Union strength in the region gleaned from a Colonel Currence, who resided in Beverly. Alexander closed the note in a defensive pose, possibly due to sensitivity caused by the Union raid on Huntersville: "I am not a military man neither am I a Politician or glory fighter. I am an humble farmer and the future will develop trooth and it will also develop falshoods as black as hell. I am sir a soldier during the war and ondly ask your attention to the facts in this matter." Alexander also noted on January 14 the arrival of Colonel William L. Jackson, who had been assigned to assume command of the post at Huntersville, adding, "I am much pleased with him as an officer."[268]

Jackson arrived in the town and discovered "everything in confusion." He assessed goods lost in the Huntersville raid as $10,227.75 in commissary stores and $2,063.66 in quartermaster's stores, and the value of privately-owned buildings destroyed as $3,000.00. Only 120 troops were present although 100 more were expected to arrive during the next few days. He bemoaned that Huntersville was "wholly indefensible with such a force, and yet I am expected to hold it. I am in the delightful position of being sacrificed if the enemy attack me with anything like the force they had when they made their attack."

He later described the town as "an awful mean place. At Allegheny we were all becoming comfortably fixed, and were living well. Here there is nothing. There are no families, my soldiers occupy the houses not burned, and the whole country has been stripped by the forces formerly here." In addition, the roads were in horrendous condition due to the thousands of troops who had marched through the town. According to one soldier, the streets of Huntersville were "simply mud holes." He added, "The fences had been destroyed; the side-walks were torn up and in a dangerous and disconnected condition."

Jackson worked at organizing the demoralized command but simultaneously sought to be relieved of the duty assigned him. He was concerned about whether the soldiers of his regiment, the 31st Virginia, would re-enlist and believed that without his presence, most would return home to "fight as guerillas."[269]

Two days after his arrival a substantial crowd gathered in the town to

Colonel William L. Jackson of the 31st Virginia Infantry. Library of Congress, Prints & Photographs Division [LC-B8171-2957]

hear a speech by a General Davis of Greenbrier County aimed at arousing Confederate sentiment. When Davis failed to appear, Colonel Jackson spoke. He informed Johnson, "I think I took the right ground with a view to your instructions," and the gathered citizens adopted a resolution recommending the formation of local companies in every neighborhood.

Jackson no doubt appreciated the sentiment but realized that with two companies from the area already serving in the Confederate army, he could expect little assistance from the residents of sparsely populated Pocahontas County except perhaps early warning of any Union advance on Huntersville. He was optimistic that forty men could be raised in the Little Levels who would prove helpful. While his scouting parties now ventured along the Huttonsville-Huntersville pike only as far as the blockade, this local force could expand operations farther north.

Jackson believed the only way to prevent a recurrence of the Union raid on Huntersville was to station two hundred men in the town and another two hundred at Greenbrier Bridge. He hoped to be reinforced in order to permit him to station a large enough force at the bridge to repel or at least delay another raid on Huntersville, where supplies were once again being accumulated. Due to the weakness of his force at present, however, Jackson planned to keep all the troops quartered in the town and to make a defensive post at the pass two miles west.[270]

Two companies of the 58th Virginia arrived in Huntersville on January 19, bringing the number of troops in the town to 245. General Edward Johnson wrote to Jackson on January 21, approving of his actions in restoring Confederate control of Huntersville and expressing his belief that the Federals in the Cheat Mountain region would not attempt another attack on the town due to the inclement winter weather. Still, Johnson encouraged Jackson to stress to his civilian scouts the importance of giving an early warning in case of a Union thrust at Huntersville and promised to reinforce him with troops and artillery in the event of an attack. He also requested that Colonel Jackson attempt to recover supplies reportedly taken by local citizens.[271]

On January 25, Jackson related his concern that another Federal raid on Huntersville was imminent. Dr. Samuel Gibson, who had led the Union forces in their attack on the town, had informed friends that a Union expedition aimed at capturing Jackson's command would be launched during the latter part of the month. Colonel Jackson expressed his view that "to demolish a small command like this one would delight the Yankees, and if the weather will permit them to do so, the attempt will be made," and he requested reinforcements and two pieces of artillery. Meetings were held this day in several neighborhoods to form local companies. The garrison at Huntersville finally was reinforced on January 30 by a section of Rice's battery and part of the 44th Virginia Infantry.

Their arrival made a food shortage caused by bad roads even worse, and for several days the men received only salt beef. Finally shipments of flour, sugar, coffee and molasses began to arrive, supplementing their meager meals. Although the shortage of food and supplies had been resolved, discipline problems, as well as an excess of one item—whiskey—were of concern. In a dispatch to Jackson, General Edward Johnson wrote, "Let the insubordinate members of your command know that they will be sent here and severely dealt with for violation of orders. Destroy all the whiskey you can lay your hands on."[272]

Whiskey was blamed for a deadly crime perpetrated by a few "desperadoes" who had been hanging around the Confederate outpost at Huntersville. On January 24, after consuming a fair amount of the drink, these men headed up Knapp's Creek and abducted two locals—Henry Arbogast and Eli Buzzard—from their homes, informing them they were to be taken to Huntersville to answer to charges of being Union men. After traveling a short distance, however, the outlaws "shot them down like dogs, and left their lifeless bodies lying in the road." These men were not actively participating in the conflict, and their murders caused great fear in the hearts of those living in Pocahontas County.[273]

Disaster also struck Confederate supporters. Between two and three in the morning on January 25, stables owned by M. B. Gillilan of Mill Point were consumed by fire. Gillilan lost five horses, quantities of various grains, and much farm equipment. His total loss was valued at more than $1,800. The fire was believed to be arson, ignited by Union sympathizers from the area. Gillilan had served as an assistant quartermaster during Loring's stay in the area in the summer and early fall of the previous year. The *Staunton Spectator* noted, "We hope the citizens of that section will exercise sufficient vigilance to detect some of these vile rascals and succeed in having justice served to them."[274]

On Allegheny Summit, the dull routine produced tension that occasionally erupted into violence. A feud between two companies of the 12th Georgia grew over a period of days and finally "found vent" on the evening of January 18. A fight between two "drunken Irishmen" and a member of another company escalated, resulting in "the wildest confusion." One Georgian who emerged from his tent observed, "Men were throwing and flourishing knives, bleeding and swearing. It was with difficulty that quiet was restored. One man was badly cut and struck with rocks."[275]

For the most part, however, the routine of camp life was predictable, broken only by false alarms. Work began on a stable large enough to house 150 horses. Capitalism remained alive and well at Camp Allegheny as revealed by Private James Hall and some of his comrades, who went into the pie business, baking about a dozen per day and selling them for $3.25 apiece.

Four inches of snow fell on January 24. "It stormed and blowed like the Dickens here," wrote Adam Kersh, and "blowed the snow through our old clapboard roofs and sifted it over us some." The soldiers had moved into their cabins on January 10, but many would have preferred to remain in their cotton tents with attached chimneys, "which render them very comfortable." An alarm sounded during the day when the sound of gunfire was heard in the rear of the Confederate camp, but it was soon discovered that the ruckus had been caused by a party getting boards, who had fired their weapons.

Fighting erupted on January 27, but the weapons were snowballs rather than guns or knives, as the 52nd Virginia and the 12th Georgia battled for several days. On January 30, the Confederate pickets stationed at the old church reported firing, which alarmed the camp and broke up prayer meetings. The alarm turned out to be false. A few moments later, a sentinel at the church fired his weapon, which led the brigade to form for battle. Again, however, it was a false alarm. One soldier noted, "we have some very good times here although we have to run out in the ditches sometimes when the pickets makes a false alarm. I tell you that we get up and toddle to the ditches and there we have to stand out there and all most freeze, but we take it all in fun."[276]

Five inches of snow fell on February 2. On the following day, Adam Kersh of the 52nd Virginia received a letter from his brother informing him of the death of their mother. Kersh replied, "The thoughts of never seeing the object I held most dear on earth makes me feel very sad indeed. But we have the consoling thought that she is gone to a better world where the wicked cease from trouble and the weary are at rest. Therefore we ought to try and walk more upright and try to obtain that rest which is prepared for the righteous in a better world than this that we may meet together in that land where parting is no more."[277]

Daily life was also difficult for the soldiers' loved ones at home, who had to carry on without their sons, brothers, fathers and husbands. Mary Dedrick of Augusta County worried about the temptations faced by her husband Henry, who served in the 52nd Virginia. She also expressed her concern about his prospects for eternal life, writing, "Uncle Jonathan says watch and pray lest you be led into temptation for he says your wife is here and you are there. Dear Henry I am glad that you have not give up trying to get to heaven. In this world we have tribulation. But in Christ we have consolation. I hope we will meet around the throne one day or other. Dear Henry strive for heaven."[278]

On January 27, Confederate Captain C. G. Barksdale, who headed Major George Jackson's scouts, rode to Dry Fork to arrest a Union man reported to be in that region. Their target had fled, but Barksdale discovered that a body of Federal troops was moving toward the area and sent a dispatch to Captain Joseph H. Lantz that evening requesting that his company move immediately toward the Dry Fork. Lantz did not respond. Barksdale and a few men remained in the area and on January 29 he saw five Union soldiers, which led him to again request Lantz's assistance. About a dozen of Lantz's men volunteered and, led by a 4th sergeant, began the trek to the Dry Fork. Barksdale had designated a meeting place, but before Lantz's men reached this point, three of them determined to scout a side road, stating that they would report by eleven the following day if they discovered Union troops. If not, they would follow the road back toward Lantz's camp without making further contact.

The remainder of Lantz's men linked up with Barksdale and scouted the road along the Dry Fork. On the following day, after waiting until three in the afternoon, Barksdale and the small cavalry detachment withdrew from the Dry Fork region and rode toward headquarters at Harper's in Pendleton County. They passed Lantz's camp late at night and didn't stop;

consequently, they failed to learn that the three men who had ventured off on their own had been captured by Union soldiers.[279]

On February 8, Major George Jackson ordered Captain David Fluornoy to lead a scout to the Dry Fork and became uneasy when he had not returned four days later. Jackson learned on February 12 that Moorefield had been taken by the enemy. Upon the approach of the Union force the militia occupying the town "made a hasty and inglorious retreat."

The following day, Jackson prepared to send an additional party, but Fluornoy's men returned that evening without their leader, who remained in the Seneca country. The scouts reported that some of the Dixie Boys, including Captain Elzey, had "gone over to the Yankees." Jackson was angered and suggested to Johnson that "action should be taken in that country immediately."

At least two of the three men of Lantz's company captured on January 29 returned within days. One of them, David Simmons, was held by Major George Jackson after he returned to the region following his release. Jackson described Simmons as "a half witted fellow" and did not believe he posed a threat, but he kept him in custody temporarily because he believed it could be dangerous to allow him to return to the Dry Fork.[280]

Another of the men captured by the Union soldiers, Joseph H. Ebbon, was debriefed by Jackson. He reported that four Union pickets had surrounded and captured the three men of Lantz's detachment shortly after they had split off from Barksdale's party. On Friday, Ebbon was taken to Pennington's, where twenty Union soldiers were encamped. Twenty more Federals arrived on Tuesday. The prisoners were later taken down the Black Fork, a tributary of the Cheat River located in Tucker County, where another Union detachment was stationed, estimated by Ebbon as two hundred strong. The soldiers were scattered about the area in the homes of local citizens. A large number occupied a log house at Pennington's, where they had knocked out the chinking and even made preparations to fight out of an outhouse if necessary.

A Sergeant Stevens, who signed as captain, was in command of the scouting party. The Federals had no artillery but were well-armed with muskets, sidearms, and "an abundance of ammunition." Eight men escorted the prisoners to Beverly, where they apparently signed an oath of allegiance, although this is uncertain. They did receive a pass through Federal lines.

Ebbon informed Jackson that "all of the citizens in the country thro which we passed seemed to be Union men tho by compulsion I thought."[281]

The ten men sent out by Lantz returned with horses that had been left in the pastures at Jacob Harper's. Lantz informed Captain Barksdale that his command was too weak to attempt operations against the Union scouting parties in the Dry Fork region. Another cavalry company under Captain McDonald had been ordered to Moorefield, leaving Lantz with about twenty-five men. This could explain his failure to send troops to Barksdale immediately. Lantz implored Barksdale to send a dispatch to Allegheny Summit requesting fifty to seventy-five volunteers "to go out and clean up the lower end of Dry Fork with fire and sword and let the Penningtons hunt for a shelter." He also feared for the safety of his command. The Union presence seemed to be aimed at control of the passes through the Alleghenies; Major Jackson believed that the only way to compel their withdrawal was if Johnson sent a force down from the summit.[282]

After receiving news of the Union presence on the Dry Fork, Jackson ordered Barksdale to report to General Edward Johnson "the exact state of affairs on Seneca." In a letter to the general's adjutant, Lieutenant Edward Willis, Jackson noted that the Dixie Boys, who were principally from Randolph County, were dissatisfied with their lot, leading him to doubt he could properly organize them.[283]

The Federal troops faced their own challenges. Union General Robert H. Milroy issued reports to his superior, General William S. Rosecrans, on January 31, detailing the conditions and manpower of the four Federal posts in the Cheat Mountain region. Beverly was occupied by the 32nd Ohio and one company of cavalry, but the town had no earthworks or defenses and no artillery. The troops were scattered to command the roads entering the town and Milroy expressed his opinion that two additional regiments would be needed to resist an attack and protect about half a million dollars in supplies which were stored at Beverly.

Huttonsville was also without defensive works. Two six-pound brass cannon were on hand, but with no artillerists, they were of limited benefit. Only 200 soldiers were stationed at this post, and they were scattered in order to provide suitable winter quarters. Milroy recommended that a regiment reinforce the post but noted that it was in fact indefensible against even a small enemy force.

In contrast, Camp Elkwater, seven miles south of Huttonsville, was fortified with earthworks and occupied by 500 troops of the 25th Ohio. Four six-pounders were mounted at this post, but again, there were no artillerists present to man the guns. For the time being, a company of the Ohio regiment was being trained in their operation. Milroy believed another infantry regiment and an artillery company would make this a very strong post.

At Cheat Mountain Summit, six pieces of artillery were manned by one company of artillery. Also stationed at the fort was the 2nd (West) Virginia, 780 strong. The defensive works were most impressive, but Milroy was uncertain as to the feasibility of defending the camp. The reason, he wrote, was "that the works were intended originally to be occupied by a much larger force—and consequently are very extensive—and so situated as to be almost two distinct fortifications, the possession of either of which by the enemy would render the taking of the other an easy task."[284]

In a letter to his wife written in early February 1862, Milroy speculated about a possible move to another theater of the war. Although he detested western Virginia, Milroy was disturbed to think he would not have another chance to dislodge the Confederates on Allegheny Summit. He attributed the failure to attempt another assault to the small numbers of Union troops in the region but believed that with two or three more regiments the Confederate camp could be taken. Milroy ridiculed the "setting hen" tactics employed by McClellan, which left hundreds of regiments idle, some of which could have been sent to western Virginia. Milroy added, "I will keep striking the rebles so and as soon as the weather gets better I will try Alligany Camp again and clear the country to Staunton."[285]

Milroy sought his removal from the region through political connections. Congressman Schuyler Colfax of Indiana bombarded General George B. McClellan with requests to have the general sent with his former regiment to Kentucky and he had apparently gained the acquiescence of President Lincoln. McClellan initially replied, "I think I have heard of you all the 6 times. The matter will be attended to as soon as time will permit." A short time later McClellan denied Colfax's request, citing the fact that Milroy could not be removed from the region since he was the only general officer present.[286]

During the month of February, Francis H. Pierpont, governor of the Restored Government of Virginia, received a number of letters from Hardy, Hampshire, and Pocahontas counties that revealed the frightful circumstances of Union supporters in that region. One letter from Huttonsville was penned by a man who had just traveled through Pocahontas County. He reported that the rebel force at Huntersville had murdered seven Union citizens during the past month and threatened several others. He learned that court had been held in that town in the middle of February and county magistrates had levied a tax on the citizens of Pocahontas to pay for runaway slaves. Furthermore, the draft was bitterly opposed by many.

News of murder in Pocahontas County was not unfounded. Colonel William L. Jackson reported that William Skeen, who served as the Commonwealth Attorney in Pocahontas County, arrived in Huntersville on February 10 to investigate the murder of Henry Arbogast and others. A guerilla by the name of Jack Chewning, whom Jackson had intended to detach to Johnson's command to capture John Slayton, was implicated and sought for the killings, but Chewning escaped before he could be brought to justice. The body of another man named Alderman was found on February 9. He had been shot through the head. Evidence implicated two of Captain Alexander's Tennessee cavalry, three or four citizens, and Jackson's most invaluable scout, Mr. Wood, who was arrested on February 11. One of the accused escaped, but Jackson ordered Alexander to arrest the two men from his company charged in Alderman's murder.[287]

Colonel Jackson reported on affairs in the Huntersville area in a February 11 dispatch to General Edward Johnson. Johnson had recently visited the town, and Jackson's command busily continued the task of fortifying and strengthening their position following the general's departure and also carried out skirmish drills when weather permitted. Discipline had been further improved by Jackson's order forbidding the sale of liquor within ten miles of the town. He had been ably assisted in his endeavors by Major A. C. Jones, whom Jackson consistently praised, possibly in the hope that he would be relieved from command of the Huntersville post and returned to his regiment.[288]

On the following morning, Jackson received intelligence that fourteen Union cavalrymen had spent the previous night at Gibson's. He organized thirty men to proceed to Gay's, where they would be reinforced by twenty local militia. As they prepared to depart from Huntersville, Captain George E. Booker and five companies of the 58th Virginia marched into the town and reported that at least one thousand Federal soldiers were moving down the Green Bank road toward Jackson's command. The Confederates prepared for an attack, but after a long wait and a brief reconnaissance, Jackson concluded that "we have been hoaxed." During Johnson's visit, the general had commented to Jackson that "it would be well to have a false alarm occasionally," but Jackson assured Johnson that the alarm was not a consequence of this remark.[289]

Major George Jackson reported on February 17 that he had recently captured a considerable amount of Union property, including colts, cows, and calves, which he sold to local citizens and to the quartermaster at Monterey. The proceeds totaled about $800, which was turned over to the acting assistant quartermaster of Jackson's cavalry command. Jackson also reported that Captain David Fluornoy was scouting the Seneca country in the direction of the Dry Fork of Cheat.[290]

About 300 Federal soldiers of the 25th and 32nd Ohio were in the Dry Fork region, scouting through Randolph and Tucker counties. Lieutenant B. W. Blandy wrote, "A Home Guard company of twenty-seven men, on the Dry Fork of Cheat River, gave themselves up on the approach of our forces. There have been some guerilla bands and horse-thieves prowling about, and they have annoyed both the Government and loyal citizens very much. It is now determined to annihilate them. Our men have now been out thirty days, and they have the privilege of remaining out 30 more, if necessary to perfect their work."[291]

Other than these scouting parties, the Union troops stationed in the Cheat Mountain District spent most of their time attempting to deal with the frigid winter weather. They did have cause for an occasional celebration, such as the capture of Fort Donelson. The evening of February 22 was "a night of great rejoicing" in the town of Beverly, occupied by the 32nd Ohio. A huge bonfire burned in the town square, speeches were given, and nearly every house in the town was brightly illuminated. It was "a glorious time" for the Federals, but some of the townspeople were less than thrilled with the news of the Confederate defeat.

On the following day, one of the Union soldiers wrote, "There was one man in town who would not illuminate his house last night but some of the boys went to his house and told him to strike up the lights or they would tear down his house so he put out his lights." The troops at Cheat Summit celebrated by firing their artillery; the noise was heard in McDowell by the citizens, who supposed the Federals were marking the birthday of George Washington.[292]

An alarm was sounded at Camp Allegheny about 2 a.m. on February 22, forcing the Confederate troops to double-quick to the trenches, which were half-filled with snow. Many of the men marched in place to keep their feet from freezing. By daylight, their officers discerned that no attack was forthcoming, a realization that pleased at least one Georgian, who wrote, "I felt good to think wee did not have to fight. I dont want to fight as long as wee can do as well without it. I dont want to be shot at: some of them stand a good chance to hit a fellow."

The alarm had apparently been caused by scouts firing upon one another, but the infamous John Slayton appeared at Wilfong's where he took one Confederate prisoner. Scouting parties searched as far as Hightown but were unsuccessful in their attempt to capture the Union scout. Slayton's work for the Union army led Johnson to offer five hundred dollars and a discharge from the army to any Confederate soldier "that would bring old slaten to him dead or alive."[293]

General Johnson was apparently away from Camp Allegheny during the latter part of February, and, in his absence, at least two colonels assumed temporary command. Colonel Z. T. Conner, who commanded the 12th Georgia, headed the brigade on February 23, as indicated by his issuance of General Order Number 4, which called for increased vigilance, tighter security measures, and quick response to future alarms.

On February 27, in response to the frequent alarms, Colonel William L. Jackson, who had apparently assumed command of the brigade, issued General Order Number 5, which scheduled a muster and inspection for the following morning. The order also contained a warning to soldiers of the brigade who had made a habit of "flanking the pickets, & passing the chain of outposts" during unauthorized excursions from the camp and one to officers which stated that allowing their men to leave camp without proper authorization was "a gross dereliction of duty."[294]

That evening, the brigade was yet again forced into the trenches after the sound of pickets firing was heard. "We were marched to the ditches tonight about 7 o'clock and remained until eleven," one soldier wrote. "The trenches had lots of snow in them, the ground frozen perfectly hard, and the wind singing like minie balls. After we learned no enemy was coming we marched back to our quarters almost chilled to death."

Apparently two Confederate soldiers had ventured down the turnpike in search of whiskey. One of the men became extremely intoxicated, and his companion had to drag him back toward camp. They were fired on by the pickets, and the drunken soldier responded by firing five shots, forcing them back toward camp. Scouts sent from Camp Allegheny to see if the enemy was in the vicinity found the intoxicated soldier at Wilfong's where he had consumed some dinner. He was arrested and placed in the guardhouse, another victim of the spirits that were readily available in the camp's vicinity.

This incident came soon after General Johnson resorted to sorting through incoming supply wagons box by box in an effort to stem the influx of alcohol. If whiskey was found anywhere in a wagon, the entire load was confiscated, depriving many innocent soldiers of delicacies from home.[295]

In late February, Captain R. D. Lilley, who headed Company C of the 25th Virginia, wrote an interesting note to Colonel William L. Jackson, his brigade commander, which is revealing regarding the fortifications at Camp Allegheny:

> *As officer of the day I feel it my duty to report to you the condition of several of the wigwams for the protection of the pickets. The first one on the "Varner trail" has been destroyed & the pickets are of course very much exposed. The one on the Greenbrier road, near Wilfong's house, needs to be repaired. There are others that would be very much improved by having more clay placed upon them, as the clay which was first put on has settled, leaving the upper portion of the wigwams exposed to the snow, rain, &c.*
>
> *I would also respectfully call your attention to the condition of the entrenchments, as there are some improvements needed, at once, upon the southwestern portion of the ditch, near the blockhouse or battery. As, in case of an attack, the portion of the ditch immediately in front of the battery would have to be vacated, would it not be well to change the road so as to bring it up through the ditch at*

the point mentioned, & cut out the ditch through the present road, as that would save shifting the men in time of action & at the same time give the artillery better range. I would also respectfully suggest that logs be placed upon the front of the entrenchment—first place a log on the top of the bank & parallel with it, then place a log at the end of the first & perpendicular to it, allowing the ends to extend out upon the bank in rear of the ditch, after which place a log on top of the two last & parallel with the first. This would form a strong breast-work, the timbers being heavy, and the space between the logs would answer for the men to fire through, while the top log would furnish a great protection to the heads of the men.[296]

CHAPTER EIGHT
CLOSING THE SHOW

Throughout the month of March 1862, Federal scouting parties continually probed the Confederate forces in the region, as they prepared to advance eastward along the Staunton-Parkersburg Turnpike. The month would be filled with skirmishes and plans initiated by both sides to go on the offensive as the season prepared to change from winter to spring.

In the early part of March, Lieutenant A. J. Weaver led sixty-three picked men of the 2nd (West) Virginia Infantry on a mission to destroy Confederate supplies stored at Green Bank. Weaver's detachment made their way through the mountains and reached the town at about two in the morning. Upon reaching the outskirts of Green Bank, Weaver ordered Sergeant George Quimby of Company K with ten men to proceed into the community to reconnoiter and to return in half an hour. Quimby's detail stopped at the first residence they saw and were informed that six Confederates were sleeping in the church. The Union men surrounded the church but upon entering the sanctuary found no enemy soldiers. They exited the building and entered the courtyard when they spied a group of men coming down the street toward them. Quimby and his men knelt behind a fence and prepared to fire. As the sergeant began to issue the command, the Federal troops recognized the voice of one of the approaching soldiers as Lieutenant Weaver, their commanding officer. A tragic incident had been narrowly averted. After burning a building filled with supplies, the Union party returned to camp sometime after dark.[297]

Another Federal expedition headed by Captain Sylvester Phillips of the 3rd (West) Virginia Infantry was launched into Webster County in early March. One hundred infantry and some cavalry snagged four prisoners including a man named Starks, a "noted hors thief." One of his captors struck up a conversation with the infamous Starks and discovered he was the son of a shoemaker he knew from his childhood. This soldier also observed that "the Webster boys are a wild set they seem to be an ignorant set." A number of men from the area visited the Federal camp at Addison, and an officer delivered a speech calling on them to volunteer as scouts. Webster was described by one soldier as "a land flowing with shugare water and honey," and the Union troops took advantage of the bounty when opportunity arose. One wrote, "I call it stealing perhaps some find a milder name for it." This same soldier noted, "there are some farms there that I think are 50 or 60 years old which were first settled by hunters."

During their stay in Webster County, one of the soldiers ran after a chicken and ventured about one hundred yards from camp. Suddenly rifle shots rang out. One of the bullets nicked the top of a Union soldier's pants and blasted chips from a sycamore tree into another soldier's face. Another bullet ripped into one of the Federals, who fell wounded. Private Matt Cutright turned around and "told them to shoot again then he came to the house. Some 5 or six of the boys sieced their guns ran through the river knee deep and up the hill through the woods but did not get one of them some of the cavalry shot at them but with out effect."

The next day the Union troops killed one enemy combatant, but upon leaving the area they were forced to leave their wounded comrade behind. A member of the 3rd recorded, "I understand he has since been shot."[298]

There was also activity in early March in the Dry Fork region. During the last week of February, General Robert H. Milroy had ordered a two-pronged Federal expedition into the rugged and isolated area. One hundred and fifty troops of the 32nd Ohio departed from Beverly while Captain George Latham led two hundred soldiers of the 2nd (West) Virginia from Cheat Summit to the North Fork of Cheat River. Both Union parties, battling rain and snow, pushed through the mountains, frequently forced to wade frigid, waist-deep streams. The two scouting parties rendezvoused at the Dry Fork of Cheat on the last day of February and proceeded cautiously in search of the enemy.

On March 1, a little more than a mile from the Mouth of Seneca on the Black Fork of Cheat, they discovered a company of Confederate Home Guards, formed and led by Captain Joseph H. Lantz, encamped in an old Lutheran log church renamed "Camp Luther." Upon the approach of the Union force, Lantz's men fled "like wild geese" but not before Lantz and at least ten of his men were taken prisoner.

On the following day, part of Latham's command moved up the North Fork. As they neared Stringtown, they were fired on by the Dixie Boys and a small Confederate cavalry force which had gathered in a narrow defile to resist the Union invaders. Although the Federals were temporarily thrown into confusion, the Confederate cavalry fled, allowing the Union force to concentrate their fire on the Dixie Boys, concealed along the mountainside. Outnumbered, the Confederates were forced back and finally withdrew, leaving two dead comrades, whose bodies were dragged down the mountain and flung across a fence.

The captured Confederates were sent to Columbus, despite the claims of some that they had been forced into service. Milroy claimed his men killed three, wounded eight, and took 28 prisoners and 58 horses without losing a man. Following the return of Latham's force, another expedition was launched into Webster and Pocahontas counties, aimed at quelling the armed resistance of the citizenry. The force of 300 was led by a member of General Milroy's staff; little more is known about this scout.[299]

While Confederate Captain James C. Arbogast of the 31st Virginia carried the news of the Union expedition to General Edward Johnson on Allegheny, Major George Jackson, who commanded the Army of the Northwest's small cavalry unit, headed into the Seneca country to ascertain the strength of the Federal force. He returned on March 3 and reported the enemy present in considerable numbers on the North Fork and that the destruction of Lantz's command eliminated the possibility of a cooperative effort to drive out the enemy. The Dixie Boys were "scattered about in the wood without any organization," Jackson noted, reporting that several had joined his command. He stated his belief that the Union detachment would encamp at Circleville and suggested that Johnson detach a force from Allegheny Summit to drive them from the North Fork.[300]

Johnson responded by ordering the 31st Virginia, with at least one piece of artillery, to depart for Crab Bottom at noon on March 4. This regiment arrived in the village that evening and encamped with the 44th Virginia. On the following day both regiments moved out, arriving in the Pendleton County seat of Franklin late in the evening. However, the Union party they expected to find in the area had learned of the enemy's approach and withdrawn to the Dry Fork of Cheat. The disappointed Confederates returned to their camps, the 31st Virginia arriving at Camp Allegheny on March 8. On this same day, a Union soldier on horseback was shot and killed "out in the country."[301]

On March 7, Jackson ordered Captain David Flournoy and eighteen of his cavalry company to reconnoiter the North Fork and Mouth of Seneca regions, where another large body of Union troops had been sighted. After crossing North Mountain, Flournoy began making inquiries of local citizens, whose reports led him to believe that at least one hundred Union soldiers were stationed at or near the Mouth of Seneca. Fluornoy's party encountered no enemy, but Union parties had been spotted by local citizens on the previous two days. He decided against attacking the Federals and instead recrossed North Mountain and returned to Franklin at midnight.

Jackson informed Johnson's headquarters that while most of the Union force estimated at about five hundred had departed from the region, small parties were still spotted. Some of the Union soldiers informed local citizens that they were returning to Beverly to obtain their pay and that they would soon return with reinforcements to move against Franklin and Circleville. Jackson again requested assistance, preferably a force to move into the Seneca and North Fork country via Circleville.[302]

Jackson commended the column commanded by his father, Colonel William L. Jackson, that had been ordered to Franklin. The soldiers under the elder Jackson held meetings and organized home guards for the region's defense, and according to his son, this "has had a fine effect upon the citizens of this country." On March 11, Major Jackson reported that enemy soldiers were still seen in small parties in the passes and woods on the North Fork. His intention to muster the Dixie Boys into service was prevented by the raid on Lantz's camp, which left only thirty men ready to join the Confederate army. The men of the company believed they could raise a few more volunteers, and Jackson accommodated their request to wait a week before mustering them into service.

The Dixie Boys gathered again on March 15, but much to Jackson's consternation, only sixteen men appeared. Some of the prospective troops had reportedly decided to carry on as guerillas. The company did pledge to raise thirty men and were given still more time to meet this goal.[303]

Major Jackson also sought to curb efforts by local citizens to avoid compliance with the militia law. On the night of March 14, the major, who was stationed in Franklin, learned a party of citizens from Augusta and Rockingham counties were seeking to flee toward Greenland for that purpose. His cavalry company was dispatched in pursuit and succeeded in capturing seven men, who were sent to Monterey. The citizens of

Petersburg were credited with snaring more than sixty others. Many of the prisoners professed to be Dunkards and claimed that their religious practices forbade military service.

Another local company formed from men of the Little Levels region was headed by Captain Jacob Williamson Marshall. Marshall's force guarded a section of the Huttonsville-Huntersville pike that had been patrolled by two cavalry companies in the fall. In mid-March, Marshall requested that his company be allowed to remain stationed in that area, a plea endorsed by Major A. C. Jones, commanding at Huntersville.[304]

While Confederate officers attempted to enforce the militia law in the region, Union troops received news that Federal forces had taken Manassas. The 32nd Ohio, stationed at Beverly, received word by telegraph on March 11. The regimental band played a number of patriotic tunes, the courthouse bell was rung, and the men of the regiment "gave three long and loud cheers in exaltation over the victory." One member of the 32nd dined with some locals and enjoyed a supper of buckwheat cakes and molasses, along with some maple taffy.

Their high spirits soon disappeared as the regiment was ordered to Cheat Mountain on March 14. Along the march, Charles E. Smith passed the gravesites of fallen comrades John Stenbeck and Jesse Gardner. Smith leaned over the graves and "asked God to bless me and enable me to live a whole and Christian life. It was a solemn thought. I bid farewell to their graves and after picking a twig from each of the pines that grew over them, I started on." In their cabin atop Cheat Mountain, Smith and his fellow musicians passed a resolution "to have no more swearing in the mess. A great change took place. Where swearing, card playing and profanity reigned, quietude, harmony and peace took place."[305]

On March 15, Confederate Major A. C. Jones, in command at Huntersville, reported that his favorite scout, a Mr. Wood, had just returned from an expedition to the region around Beverly and Huttonsville. Wood relayed that the Union force in Randolph County had not received reinforcements and that their efforts to enlist locals in the militia were proving unsuccessful. Jones warned Johnson about a new road constructed by the Federals which led from Cheat Summit to Jackson Moore's on the Greenbrier River. Union foraging parties of about thirty men with seven wagons traveled to Mingo Flats about once a week; Jones informed the general that he would attempt to capture them at the next opportunity.[306]

That evening, six Union soldiers, guided by either Samuel Gibson or John Slayton, went through Clover Lick to the Green Bank road about eight miles from Huntersville where they captured two Confederate supporters, Earle and Kittle. One of the Union soldiers was detached to escort them back to Cheat Summit, but as soon as the opportunity arose, the prisoners made their escape. The guard fired, striking one of the men's horses, but they escaped, reaching Gum's on the Monterey road at 8 a.m. on the following morning.

Major Jones was alarmed by the Union raids conducted through the area around Back Allegheny Mountain toward Green Bank and the Greenbrier River and recommended more frequent scouts of this region. He believed that Johnson employed Captain Arbogast's company to patrol the no-man's land at the base of Allegheny Summit and cited his fear of an accidental, deadly encounter with these scouts as the reason he had not assumed responsibility for the defense of this section of the country. Jones sent out a force on March 16 with instructions to capture the next Union foraging party that ventured to Mingo Flats and to kill Union scouts Slayton, Gibson, Young, and Gum.[307]

Meanwhile, at Camp Allegheny, preparations were made to meet the expected Federal spring offensive. Roads were blockaded, and regulations regarding visitors to the camp were issued. A scouting party moved out from the camp on March 18 searching not for enemy soldiers, but for whiskey. A considerable quantity of spirits was being moved into Camp Allegheny after dark by horses driven across the mountain from Crab Bottom to Wilfong's. Private Joseph Snider of the 31st Virginia, who had been annoyed by the drunkenness prevalent in his cabin, reported the smugglers' modus operandi to General Johnson. In response, Johnson ordered Lieutenant William P. Cooper with a squad of men to scout the region northeast of Allegheny. Cooper and his men captured a Mr. Mills along with a load of whiskey, apples, and sugar and turned the bootlegger and his goods over to the provost marshal. Johnson ordered the whiskey poured out, confiscated the supplies, and turned the horse over to the quartermaster. Mills was placed in the brigade guard house.[308]

General Robert E. Lee sent a dispatch to General Johnson on March 15, notifying him of the governor's authorization for calling out the militia of Highland, Pendleton, and Pocahontas counties. Lee requested a report detailing the strength of Johnson's command and the prospects of militia turnout. He also sought Johnson's opinion as to the best location for defending the approaches to Staunton. Johnson's adjutant, Lieutenant Edward Willis, informed Lee on March 18 that the total strength of the Army of the Northwest was just less than 4,000 soldiers, but fewer than 2,800 were present for duty. Five hundred troops and two pieces of artillery were stationed at Huntersville, 350 at Crab Bottom, 40 at Franklin, 60 at Monterey, and 1,834 men and ten pieces of artillery at Camp Allegheny.[309]

On the same day, Johnson notified Lee that the prospects for reinforcements from the militia were "by no means flattering. The people of this country are many of them, if not disloyal, at least indifferent as to which cause triumphs, and are willing to range themselves on the stronger side. The population is sparse, the country very mountainous." Although the Federal force in the region was supposed to be about the same strength as the Army of the Northwest, Johnson admitted that "little can be done save hold the enemy in check, and scarcely that if he advances with a very superior force."

Despite rumors that the Union force would soon advance eastward along the turnpike, Johnson believed that, if an advance was undertaken by the enemy, it would likely seek to outflank the strong position atop Allegheny Summit by marching via Huntersville or Moorefield or to force a Confederate withdrawal by advancing to Winchester. Johnson informed Lee that the only defensive position along the pike east of Monterey was on Shenandoah Mountain, twenty-six miles west of Staunton.[310]

Upon receiving Johnson's dispatch, Lee expressed his regret that the response of the militia was expected to be light but reiterated their importance in filling out the ranks of the Virginia regiments. He also ordered Johnson to determine the possibility of setting up a line of defense along Shenandoah Mountain.

Lee added, "I need not urge upon you the necessity of keeping yourself advised of the movements of the enemy, of using every means in your power to thwart and defeat him, of increasing your own forces, and putting your army in the best condition possible for the opening campaign."

Not only did the prospects of reinforcing the Army of the Northwest appear grim; many of the Confederates at Camp Allegheny had no plans for re-enlisting. Colonel William L. Jackson of the 31st Virginia bemoaned the desire of many to return to their homes and serve in local guerilla units, writing, "They all have the Guerilla fever. There is a wild, roving and daring life connected with that system, and certain freedom from discipline and restraint, dazzling to the most of them."[311]

As the Confederates feared, the Union command in western Virginia had indeed begun planning for a spring offensive. General William S. Rosecrans proposed to dislodge the Confederates at Camp Allegheny by moving a force of five regiments up the South Branch Valley via Franklin in conjunction with an advance by Milroy's command along the Staunton-Parkersburg Turnpike. The forces would then combine to seize the railroad at Millboro. Milroy's Cheat Mountain District was manned by slightly more than 6,000 troops. General Milroy proposed to make his advance with 4,000 men, adding, "After taking Monterey and Alleghany I should be re-enforced to make sure of holding possession at Staunton, if thought best to move on there."[312]

Milroy revealed his plans in a March 16 dispatch to General William S. Rosecrans. He proposed to march seven miles beyond Cheat Summit on an unfinished road to Green Bank and then on to Huntersville, which was lightly defended. There Milroy hoped to link up with a force moving northeast from Lewisburg led by General Jacob D. Cox. The combined armies would move on to Monterey, cutting off the Confederate force on Allegheny Summit, before marching quickly to Staunton, "or, if thought best, on to [the] Virginia and Tennessee [rail] road."

Milroy complained of a shortage of carriages for his artillery and also requested a battery of fully equipped rifled guns, as he had only smooth bore cannon. He wished to advance promptly but requested Hyman's battery and one or two additional regiments if the move east was to be permanent. While the roads between Webster, the nearest rail hub, and Huttonsville were in bad condition, the turnpike was in fairly good shape on the approach to Allegheny Summit. Milroy closed the note with a confident air: "Fear that game I have watched so long will escape me at last."[313]

Milroy stressed the necessity of striking at once due to the anxiety in Pocahontas County caused by the Confederate draft. He informed General Rosecrans that many of the county's male citizens were heading into the mountains to escape. Four days after the draft law took effect, Milroy noted that forty-six refugees from Pocahontas and Highland had come into his lines and reported that the legislation had caused "general consternation" in these counties. Milroy was seemingly convinced that an offensive through Pocahontas County would alleviate the pressure exerted by Confederate authorities on the county's populace: "They beg and pray for protection; say hundreds will assist in driving out traitors if we come over." Refugees from Pocahontas County continued to trickle into the Union camps in the following days, "fleeing from the mobs who prowl through that county attempting to enforce the order of Letcher, to draft the militia."

According to a soldier of the 3rd (West) Virginia, they "confirm the information which we have long heard, that the majority of their fellow-citizens in the old mountain county are in favor of the Union." This soldier also noted that his regiment had been "considerably augmented by recruits from among these loyal men who have fled from the tyranny of the nigger aristocracy" and that a new Union company designed for service in the 10th (West) Virginia was to be composed entirely of Pocahontas County natives. Milroy provided aid to the neediest families in the vicinity of Huttonsville and Beverly, many of the men having joined the Confederate army.[314]

In the latter part of March, Milroy dispatched a scouting party of thirty men from Company C of the 3rd (West) Virginia to reconnoiter the country around Huntersville. These troops were under the command of Sergeant Dennis B. Jeffers and were accompanied by five experienced guides hired by the general. The Federal soldiers followed the Huttonsville-Huntersville pike in a southeasterly direction. A party of about 55 Confederates who were out driving in the militia of Pocahontas County came across their trail and started in pursuit, and they reportedly sent a request for assistance to a company stationed at Clover Lick, some six miles distant.

After crossing Elk Mountain about seven miles from Huntersville, the Union detachment came to the blockade, when they discovered the Confederate force in their rear and prepared to attack them, "not dreaming of the hornets nest they were getting into."

Suddenly two Confederate companies sent to the scene by Major A. C. Jones, commanding at Huntersville, attacked the Federals in their front. A sharp skirmish erupted and the Federal scouting party quickly found itself surrounded by a superior enemy force. Gunfire came from all directions; nearly all the Union soldiers had clothing cut by bullets. Although some of the men were merely grazed, two were reportedly killed and several others wounded. A desperate effort to break out was successful, but six members of the expeditionary force, including Sergeant Jeffers, were missing when they returned to Elkwater. Two wounded Union soldiers, Privates Christopher Albright and George B. Spurgeon, were captured by the Confederates. Milroy called the escape of the Union party "almost miraculous."[315]

By the end of March, it was evident that the Confederate Army of the Northwest would be withdrawn in a short time. Only ten days' supplies were on hand; still, Johnson remained cautious in anticipation of an attack on Camp Allegheny. One soldier noted, "The camp is as well picketed as any camp can be. They cant surprise us, to save them. The guard duty is verry heavy: a man has to be on guard every 5 or 6 days."[316]

Despite the impending Confederate withdrawal, Johnson's orders indicate that he not only considered holding the line, but even pondered launching an assault on the Union post at Crouch's, just south of Huttonsville on the road to Huntersville. Major A. C. Jones supported the plan and thought the mission could be accomplished with a minimal force. Scouts reported only six companies were encamped at the site.

The Confederates planned to attack from two directions. Five or six companies from Camp Allegheny were to cross Cheat Mountain and pick up Donnelson's trail at the head of Windy Run, following this path to the right rear of the Federal camp, while three companies from Huntersville would march up the Huttonsville-Huntersville Turnpike. This contemplated offensive apparently did not progress beyond the preliminary planning stages.[317]

In response to an inquiry from General Johnson, Major A. C. Jones informed him that no quarters were available in Huntersville to house additional troops in the town. The cabins built in the summer and fall of 1861 under the direction of General W. W. Loring had been left for the most part half-completed. The few finished shelters were occupied by troops of the 58th Virginia, while many of the uncompleted cabins had been destroyed or hauled away by local citizens, probably for firewood. In fact, to accommodate the five companies of the 58th stationed in the town, Jones had been forced to erect several board sheds. Supplies were low due to the belief that an eastward withdrawal was imminent.[318]

As preparations for the Confederate withdrawal intensified, so did concerns about the reenlistment of the soldiers of the army. The Pocahontas and Highland militia entered camp on March 28; most volunteered in different companies, while the rest were mustered into the Barbour Greys. On the previous day, the Virginia units had been mustered for reenlistment, but positive responses were practically non-existent. As a result, their officers were forced to resort to harsher measures. On March 28, Colonel William L. Jackson of the 31st Virginia made a brief speech and then ordered his men to their quarters. When some refused, Jackson ordered their arrest. Several escaped, but others were captured and sent to the guardhouse.

On March 30, the regiment was again formed and the proceedings of a court martial were read to the men. Several were found guilty of desertion and sentenced to six months' hard labor on the batteries at Richmond while shackled to a 24-pound ball. General Edward Johnson remitted the hard labor due to the impracticality of the sentence but did order the forfeiture of six months' pay. One member of the 31st Virginia who had been arrested for stealing a box of tobacco was to be paraded daily before his regiment for ten days carrying a board on his back with the word "thief" inscribed in large letters.[319]

The Confederate troops at Camp Allegheny were aroused at 4 a.m. on the first day of April to load the several hundred wagons sent to move supplies eastward. Their preparations were observed by Union troops atop Cheat Summit. The main body of the Army of the Northwest left their winter camp on the following day. The withdrawal of the troops "caused a great panic in Highland, Bath, and Pendleton counties. Many of the people were flying to get away from the Yankees. It was really painful to witness the anxiety of the women." The Confederate soldiers at Huntersville also moved east.[320]

On March 29, Confederate President Jefferson Davis declared martial law in a number of western Virginia counties including Pocahontas and Randolph. In Pocahontas County, citizens lived in fear of the two companies of Virginia State Rangers, commanded by Captains George Downs and John L. Spriggs, which had been authorized by the General Assembly on March 27, 1862.

On April 4, Commonwealth Attorney William Skeen wrote to Confederate General Henry Heth, commanding the District of Lewisburg, and demanded that he disarm and disband "these lawless banditti," who were responsible for the murder of three citizens of the county and the theft of fifteen to twenty horses. "They are supreme judges of the loyalty of Virginians," Skeen fumed, "and pass sentence of death or confiscation of property without evidence or the shadow of it, and execute their sentence of death and decretal orders of confiscation on any man they desire to kill and rob, or who has the misfortune to have sufficient property to be styled by them 'Union men'."

The militia had been called out, leaving their homes and families unguarded. Skeen urged that these men be sent back to Pocahontas County to provide protection for area residents and their property. Heth immediately wrote to Virginia Governor John Letcher, informing him that he intended to disarm the companies of rangers commanded by Downs and Spriggs. Heth also requested that the governor refrain from forming other similar organizations in the region.[321]

According to one soldier, an even more horrible atrocity was committed by Union troops. On April 1, after heading about four miles from camp at the site of an earlier encounter with the enemy, a fifer from the 2nd (West) Virginia and a soldier from Company A of the 14th Indiana dug up the bodies of two Confederate dead, decapitated them with a shovel and placed the bodies in a camp kettle where they were boiled until the flesh parted from the bone. Later, other soldiers stationed here boiled pork in the same kettle but hastily rid themselves of its contents upon learning of its earlier use. The Union troops were enraged, calling the men "The Resurrectionists grave robbers and other names. We had a sing in the evening." No further information was found to confirm this perverse tale.[322]

The long-awaited Union advance along the Staunton-Parkersburg Turnpike would commence during the first week of April 1862. Milroy wrote to Major General John C. Fremont on March 31, updating him on affairs in his district:

> *Refugees continue to come in squads of from 5 to 25, in great destitution. Some have enlisted in Virginia [Union] regiments and some employed by me on roads. Twelve, arrived this day from Pocahontas, report that impressment is continuing, and assure me of strong Union feeling in Pocahontas, Highland, Greenbrier, Bath, Alleghany, and Rockingham Counties. They implore our*

> *protection and pray for assistance; was on point of giving them relief two weeks since, but in obedience to orders I deferred it. Three fugitive slaves from Highland, just in, state 80 wagons passed Monterey last Friday for Camp Alleghany, and heard their master, a Colonel Kincaid, say that enemy were going to move. Sent scouts to watch movements. Telegraph between this place [Huttonsville] and Cheat Top not working. Last week, guerillas, 300, attacked Union settlement in Pendleton; were repelled by 75 Union citizens; were re-enforced and drove back the citizens.*

Security was tight as Milroy's preparations for an advance neared completion. Two women rode into camp, wishing to pay for some horses reportedly captured by Union soldiers. John Slayton reported them as secessionist sympathizers, and they not only failed to procure the horses in question, but also lost the mounts they were riding and were prevented from leaving the camp.

On April 4, twenty-one local citizens appeared at Cheat Summit Camp, headed by a horseman who was waving a white flag. They were quickly surrounded by curious Union soldiers. The citizens informed the soldiers that "they had lain out in the woods a long time and dared not go home only at the dead hour of night and then right away again to keep away from the hands of the Rebs. They fled from their homes, left their families to save their lives. Said a band of guerrillas was scouting about harassed the Union men and stealing their property."[323]

As the Federal troops prepared to advance, scouting parties were dispatched to test the waters. In response to the activity in Pendleton County, 300 troops of the 25th Ohio, under the command of Major George Webster, departed from Beverly and moved in the direction of Franklin via the Mouth of Seneca, while a scouting party of 90 men of the 2nd (West) Virginia, led by Captain George Latham, marched toward Monterey via Circleville and Crab Bottom. On April 3, John Slayton led twenty men to reconnoiter Camp Allegheny. They found the place deserted and spent the night amidst the impressive fortifications.

The main Union force under General Robert Milroy left Cheat Summit at 1:00 p.m. on April 5 and camped for the night along the Greenbrier, opposite Camp Bartow. After breakfast, the troops marched through the fortifications and began the climb up Allegheny Mountain. About noon they entered Camp Allegheny and planted the national flag on top of the fortifications.

One soldier noted, "The camp is situated on two hills and several forts command the vicinity in every direction. The quarters are substantial log buildings, well arranged and streets and alleys between them."

Upon his arrival, General Milroy was complimentary of the Confederate defensive position:

> *I was greatly surprised at the great extent and immense strength of the reble fortifications and winter quarters here. The greater part of these works have been constructed since our fight here on the 13th of Dec. last. We must have given them a tremendous scare at that time for their whole force seems at that time worked on the fortifications since then, and it is wonderful to see the amount of labor they have performed. The place is now almost a Gibralter for strength both by nature and art, and with a brave garrison, a good cause and sufficient supplies it would be impregnable to almost any force that could be brought against it.*

As the Union troops surveyed the battleground of December 13, they discovered unburied the bodies of Lieutenant Alfred Sickman of the 2nd (West) Virginia and Corporal Benjamin F. Huntingden of the 9th Indiana. Milroy noted, "They were much decayed and were only recognized by the spots where they were known to have fallen."

The Federals also found the Confederate cabins and stables still standing and a great deal of forage which had been left behind. The widow Yeager informed Milroy that the buildings had been left standing as a favor to her and asked the Union general to extend the same courtesy. "It is said that [General Edward] Johnson is a widower and has been courting her," Milroy wrote, "This is probably the reason why the quarters have not been burned." According to Evelyn Yeager, a few Union men, including John Slayton, set fire to every other cabin the night after the Confederate withdrawal. The widow Yeager begged them to leave the buildings for her use but was told she was fortunate they did not burn the homestead.[324]

Upon learning that Latham's force had taken possession of Monterey, Milroy ordered 400 troops of the 2nd (West) Virginia under Colonel John Moss to reinforce the small detachment. He also sent the 32nd Ohio to Huntersville and dispatched a cavalry company "to scour the guerillas out

Lieutenant Alfred Sickman, 2nd (West) Virginia Infantry, killed at the Battle of Allegheny Mountain. Francis S. Reader, History of the Fifth West Virginia Cavalry (New Brighton, PA: The Author, 1890)

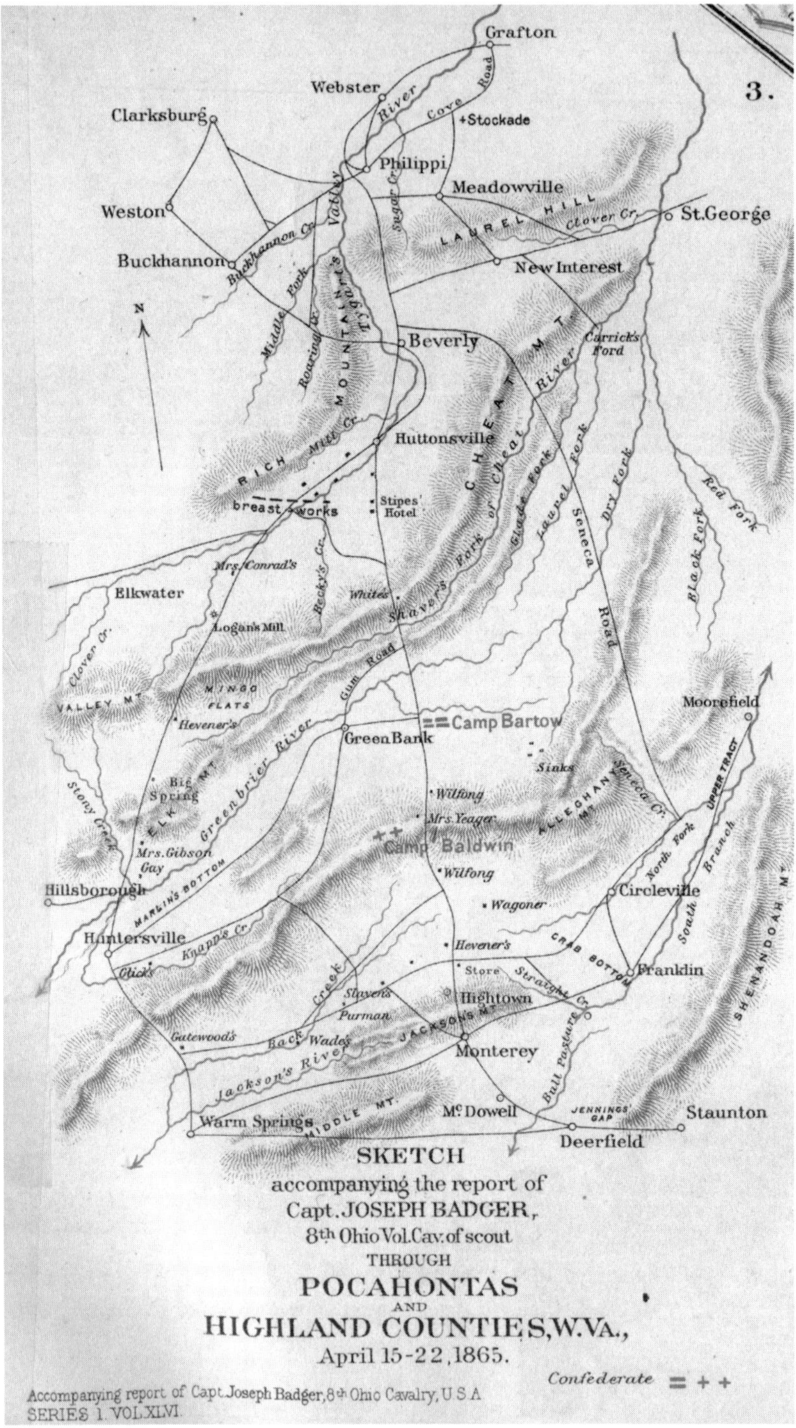

Sketch of scout through Pocahontas and Highland Counties, April 15-22, 1865. Official Records

of the country around that place." It was too late, however, for Milroy to entrap the Confederate Army of the Northwest, which had escaped to fight another day.

The lengthy stalemate between Union forces stationed in the Cheat Mountain region and Confederate forces stationed first at Camp Bartow and later at Camp Allegheny was over, and both sides prepared for the spring campaign. On May 8, the two brigades that had braved the bitter winter would meet again in battle along the Staunton-Parkersburg Turnpike at McDowell. The resultant hard-fought Confederate victory marked the beginning of General Thomas J. "Stonewall" Jackson's renowned 1862 campaign in the Shenandoah Valley.

Given the terrible weather endured by both Union and Confederate soldiers during their stay in the mountains of western Virginia, it is fitting that Milroy closed his April 7 dispatch with the words, "Terrible snow-storm here to-day."[325]

Union and Confederate armies would continue to pass over the turnpike for the rest of the war. Three years after the Confederates withdrew from their mountain fortress atop Allegheny Mountain, and less than one week after Confederate General Robert E. Lee surrendered his command at Appomattox Courthouse, Captain Joseph Badger led a detachment of the 8th Ohio Cavalry on an expedition that passed through Pocahontas and Randolph counties. He described the sentiment of the citizenry in this area of western Virginia:

> *I found the people all through my route completely conquered; Lee's surrender has finished them. They see no hope in further resistance and are willing to submit on our terms. They seemed gratified when they heard the terms. There are a great many stragglers and deserters all through the country who have no idea of going back to the army, but being afraid of us they run and hide as much as possible. I believe when they understand our terms they will all come in and give the parole. I met several paroled men of Lee's army. I judge from their talk that they will be an army of missionaries all through the South to preach to the people a doctrine of common sense and the folly of further resistance. From the spirit of submission that seems to have taken possession of the people since Lee's surrender, I believe that kindness and leniency*

toward them now would have the most beneficial results. Their only desire seems to be to get back their sons, brothers, and husbands who have been in the army, and live in peace.[326]

NOTES

INTRODUCTION

1. Ambrose Bierce, *Battlefields and Ghosts* (Palo Alto, CA: Harvest Press, 1931), 13-15.
2. John G. VanMeter, ed., *To Water Street With Affection: The Civil War letters of 2nd Lt. Charles W. Delay, Company E, 25th Virginia Infantry* (n.p.: The Author, 1999), 5.
3. VanMeter, *To Water Street With Affection*, 12; *Spirit of Democracy*, 19 February 1862; Frank Ingersoll to Sister, 28 October 1861, Frank Ingersoll Letters, Lilly Library, Indiana University; *Indianapolis Daily Journal*, 28 October 1861.
4. Charles R. Adams Jr., ed., *A Post of Honor: The Pryor Letters*, 1861-1863 (Fort Valley, GA: Garret Publications, Inc., 1989), 70; VanMeter, *To Water Street With Affection*, 12; Recollections of Christian Kuhl, Boyd B. Stutler Collection, West Virginia State Archives.
5. George R. Cryder and Stanley R. Miller, compilers, *A View From the Ranks: The Civil War Diaries of Corporal Charles E. Smith* (Delaware, OH: Delaware County Historical Society, Inc., 1999), 4.

CHAPTER ONE – WAR ERUPTS IN WESTERN VIRGINIA

6. George B. McClellan to "Soldiers of the Expedition," 26 May 1861, *United States War Department, War of the Rebellion: A Compilation of the Official Records of the Union and Confederate Armies* (Washington: Government Printing Office, 1881-1901), Series I, Volume II, 49 (hereafter cited as *OR*).
7. George B. McClellan to Winfield Scott, 27 April 1861, *OR, Series I, Volume LI, Part I*, 338; George E. Moore, *A Banner in the Hills* (New York: Appleton-Century-Crofts, 1963), 52.
8. R. E. Lee to George A. Porterfield, 4 May 1861, *OR, Series I, Volume II,* 802; Francis M. Boykin, Jr. to R. E. Lee, 10 May 1861, *Ibid.*, 827.
9. For excellent studies of the first year's fighting in western Virginia, see W. Hunter Lesser, *Rebels at the Gate: Lee and McClellan on the Front Line of a Nation Divided* (Naperville, IL: Sourcebooks, 2004) and Tim McKinney, *Robert E. Lee and the 35th Star* (Charleston, WV: Pictorial Histories Publishing Company, Inc., 1993). A thorough look at the initial campaign along the Staunton-Parkersburg Turnpike can be found in Fritz Haselberger's *Yanks From the South!* (Baltimore, MD: Past Glories Press, 1987). A brief but detailed work on the Battle of Rich Mountain is Jack Zinn's *Battle of Rich Mountain* (Parsons, WV: McClain Printing Company, 1971). Garnett's withdrawal is covered in W. Hunter Lesser's *Battle at Corrick's Ford: Confederate Disaster and Loss of a Leader* (Parsons, WV: McClain Printing Company, Inc., 1993).

10. George B. McClellan to E. D. Townsend, 7 July 1861, *OR, Series I, Volume II, 201*; E. D. Townsend to George B. McClellan, 9 July 1861, *Ibid.*, 202; George B. McClellan to Winfield Scott, 17 July 1861, *Ibid.*, 743; Winfield Scott to George B. McClellan, 18 July 1861, *Ibid.*, 743.
11. George B. McClellan to Winfield Scott, 22 July 1861, *OR, Series I, Volume II,* 752; L. Thomas to George B. McClellan, 22 July 1861, *Ibid.*, 753.
12. Memorandum by George B. McClellan, 4 August 1861, *OR, Series I, Volume V,* 7.
13. R. E. Lee to W. W. Loring, 20 July 1861, *OR, Series I, Volume II,* 986; R. E. Lee to W. W. Loring, 21 July 1861, *Ibid.*, 987.
14. W. W. Loring to George Deas, 28 July 1861, *OR, Series I, Volume II*, 1006.
15. General Order Number 10, 8 September 1861, *OR, Series I, Volume LI, Part II*, 283-84.
16. Jefferson Davis to Joseph E. Johnston, 5 September 1861, *OR, Series I, Volume V,* 829.
17. Special Order, Robert E. Lee, 9 September 1861, *OR, Series I, Volume V,* 192; For an excellent and detailed account of Lee's offensive against Cheat Mountain, see Jack Zinn, *R. E. Lee's Cheat Mountain Campaign* (Parsons, WV: McClain Printing Company, 1974).
18. The early campaign in the Kanawha Valley is detailed in Terry Lowry's *The Battle of Scary Creek* (Charleston, WV: Pictorial Histories Publishing Company, 1982). Lowry also studied the Battle of Carnifex Ferry in *September Blood: The Battle of Carnifex Ferry* (Charleston, WV: Pictorial Histories Publishing Company, 1985). The best account of the confrontation at Sewell Mountain is Tim McKinney's *Robert E. Lee at Sewell Mountain: The West Virginia Campaign* (Charleston, WV: Pictorial Histories Publishing Company, 1990).
19. Ezra J. Warner, *Generals in Blue: Lives of the Union Commanders* (Baton Rouge: Louisiana State University Press, 1964), 397-98; Isaiah B. McDonald to O. P. Morton, 19 September 1861, Hunter Lesser Collection.
20. Ezra J. Warner, *Generals in Gray: Lives of the Confederate Commanders* (Baton Rouge: Louisiana State University Press, 1959), 149-50.
21. Frank Ingersoll to Sister, 20 October 1861, Ingersoll Letters.
22. Frank Ingersoll to Sister, 28 October 1861, Ingersoll Letters; *Cincinnati Daily Commercial*, 5 October 1861.
23. John Hadley to Enos Hadley, 25 September 1861, John Hadley Manuscript, Lilly Library, Indiana University; *Cincinnati Daily Commercial*, 5 October 1861.
24. Frank Ingersoll to Sister, 20 October 1861, Ingersoll Letters.
25. Michael G. Harman to H. R. Jackson, 17 September 1861, Michael G. Harman Letters, *Valley of the Shadow: Two Communities in the American Civil War*, University of Virginia Library, 3 April 2011 (http://valley.lib.virginia.edu/papers/A6024).

26. George E. Moore, ed., "A Confederate Journal," *West Virginia History* 22, no. 4 (July 1961): 206; John Lyon Hill Diary, Virginia Historical Society; James Atkins Diary, Georgia Department of Archives and History.
27. Joseph C. Snider Diary, West Virginia and Regional History Collection, West Virginia University.
28. Hill Diary; Snider Diary.
29. *Southern Confederacy*, 6 October 1861; Atkins Diary.
30. *Southern Confederacy*, 6 October 1861; Atkins Diary; Hill Diary.
31. Hill Diary; Atkins Diary; Wes Cochran, *Pocahontas County WV 1860 Census* (Parkersburg, WV: The Author, 1989). Despite the grave danger he faced, John Slayton (listed as Slaton in census records) survived the war and is listed as 52 years of age in the 1880 census.
32. Adams, *A Post of Honor*, 68.

CHAPTER TWO – A PLAN TO PAY RESPECTS

33. W. S. Rosecrans to J. J. Reynolds, 17 September 1861, *OR, Series I, Volume V*, 603-04; W. S. Rosecrans to J. D. Cox, 19 September 1861, *OR, Series I, Volume LI, Part I*, 483.
34. John Hadley to Enos Hadley, 25 September 1861, Hadley Manuscript; William Houghton Papers, Indiana Historical Society; William Landon, ed., "The Fourteenth Indiana Regiment On Cheat Mountain: Letters to the Vincennes Sun," *Indiana Magazine of History*, December 1933, 365; Barbara A. Smith, ed., *The Civil War Letters of Col. Elijah H. C. Cavins, 14th Indiana* (Owensboro, KY: Cook-McDowell Publications, 1981), 14. The soldiers from Indiana were particularly intent on proving their bravery due to some disparaging remarks uttered by Confederate President Jefferson Davis about Indiana troops at the Battle of Buena Vista in the Mexican War. See *Zanesville Daily Courier*, 26 July 1861, and Lewis Wallace to R. Patterson, 11 July 1861, *OR, Series I, Volume II*, 676.
35. Joseph J. Reynolds to Nathan Kimball, 26 September 1861, Nathan Kimball Papers, Lilly Library, Indiana University.
36. *Cincinnati Daily Commercial*, 5 October and 11 October 1861; *Cincinnati Daily Enquirer*, 16 October 1861; *Indianapolis Daily Journal*, 8 October 1861; David Stevenson, *Indiana's Roll of Honor* (Indianapolis: The Author, 1864), 182-83; Eri Rockwell Diary, Indiana Historical Society.
37. Rockwell Diary; E. R. Wyeth Memoranda Book, Indiana Historical Society; Stevenson, *Indiana's Roll of Honor*, 184; *Cincinnati Daily Commercial*, 5 October 1861.
38. Cryder and Miller, *A View From the Ranks*, 5; Henry H. Dedrick to Mary E. A. Dedrick, 29 September 1861, Henry H. and Mary E. A. Dedrick Letters, *Valley*

of the Shadow: Two Communities in the American Civil War, University of Virginia Library, 3 April 2011 (http://valley.lib.virginia.edu/papers/A6000).

39. *Wheeling Daily Intelligencer*, 11 October 1861.
40. Ruth Woods Dayton, ed., *The Diary of a Confederate Soldier: James E. Hall* (Charleston, WV: Education Foundation, Inc., 1961), 27.
41. *Wheeling Daily Intelligencer*, 11 October 1861; Dayton, *The Diary of a Confederate Soldier*, 28; Snider Diary.
42. H. R. Jackson to D. S. Donelson, 29 September 1861, H. R. Jackson Papers, Cuyler Telamon Collection, Hargrett Library, University of Georgia.
43. Henry R. Jackson to C. L. Stevenson, 29 September 1861, H. R. Jackson Papers; H. R. Jackson to C. L. Stevenson, 1 October 1861, *Ibid.*; F. S. Bloom to John B. Baldwin, 29 September 1861, *Ibid.*
44. Henry R. Jackson to C. L. Stevenson, 1 October 1861, *Ibid.*; Warner, *Generals in Gray*, 18-19.
45. A. C. Myers to C. R. Mason, *OR, Series I, Volume V*, 888.
46. *Cincinnati Daily Gazette*, 26 October 1861.
47. *Richmond Dispatch*, 7 October 1861.
48. *Ibid.*
49. Hill Diary.
50. Atkins Diary; Hill Diary.
51. James M. King Journal, Indiana Historical Society.

CHAPTER THREE – THE BATTLE OF GREENBRIER RIVER

52. *Cincinnati Daily Enquirer*, 9 October 1861.
53. Samuel Clarke Farrar, *The Twenty-Second Pennsylvania Cavalry and the Ringgold Battalion, 1861-1865* (Pittsburgh: Published Under the Auspices of the Twenty-Second Pennsylvania Ringgold Cavalry Association, 1911), 26-27.
54. Frank Moore, ed., *The Rebellion Record: A Diary of American Events, with Documents, Narratives, Illustrative Incidents, Poetry, Etc.* (New York: G. P. Putnam, 1862), 3: 161.
55. Moore, *The Rebellion Record*, 3:161-62; *Wheeling Daily Intelligencer*, 11 October 1861; E. Z. Hays, *History of the Thirty-Second Regiment Ohio Veteran Volunteer Infantry* (Columbus: Cott and Evans, 1896), 14; *Zanesville Daily Courier*, 18 October 1861.
56. Moore, *The Rebellion Record*, 3: 162; *Wheeling Daily Intelligencer*, 11 October 1861; Catherine Merrill, *The Soldier of Indiana in the War for the Union* (Indianapolis: Merrill and Co., 1864), 92; Robert H. Milroy Papers, Volume IV, 46, Indiana Historical Society.
57. Hays, *History of the Thirty-Second Ohio*, 181-82; William Douglas Hamilton, *Recollections of a Cavalryman of the Civil War After Fifty Years, 1861-1865* (Columbus:

The F. J. Heer Printing Co., 1915), 18-19. Hamilton records that the "captured" soldiers were from the 13th Indiana.

58. Edward C. Culp, *The 25th Ohio Vet. Vol. Infantry in the War for the Union* (Topeka: G. W. Crane and Co., 1885), 31-32.
59. Samuel V. List Letter, 5 October 1861, Samuel V. List Letters, Indiana Historical Society.
60. *Barbour Democrat*, 4 September 1968; Alan Sturm, *A Partial History of the Sturm Family of Barbour County, West Virginia* (Buckhannon, WV: n.p., 1973), 9-14.
61. H. R. Jackson to C. L. Stevenson, 7 October 1861, *OR, Series I, Volume V*, 225; Warner, *Generals in Gray*, 297-98.
62. H. R. Jackson to C. L. Stevenson, 7 October 1861, *OR, Series I, Volume V*, 225.
63. H. R. Jackson to C. L. Stevenson, 7 October 1861, *Ibid.*, 224; Clayton Wilson to Ma, 17 October 1861, Hunter Lesser Collection.
64. Adams, *A Post of Honor*, 31; Irby G. Scott to Father, 5 October 1861, Irby G. Scott Letters, Perkins Library, Duke University; Recollections of Christian Kuhl; Dayton, *The Diary of A Confederate Soldier*, 29; L. M. Shumaker to H. R. Jackson, 4 October 1861, *OR, Series I, Volume V*, 234; William L. Jackson to Wife, 4 October 1861, William L. Jackson Letters, West Virginia and Regional History Collection, West Virginia University; George Washington Wooding to Pa, 4 October 1861, George Washington Wooding Letters, Mary B. Blount Library, Averett College.
65. J. J. Reynolds to L. Thomas, 4 October 1861, *OR, Series I, Volume V*, 220-21; Nathan Kimball to Joseph J. Reynolds, 4 October 1861, *Ibid.*, 221; H. R. Jackson to C. L. Stevenson, 7 October 1861, *Ibid.*, 225; William L. Jackson to Wife, 4 October 1861, William L. Jackson Letters; *Augusta Chronicle and Sentinel*, 11 October 1861.
66. *Southern Confederacy*, 17 October 1861; John Casper Branner, *Casper Branner of Virginia and His Descendants* (Stanford University, CA: Privately Published, 1913), 179; Augustus M. Van Dyke Letter, 6 October 1861, Augustus M. Van Dyke Letters, Indiana Historical Society; Frank Ingersoll to Sister, 8 October 1861, Ingersoll Letters; *Spirit of Democracy*, 6 November 1861.
67. *Indianapolis Daily Journal*, 14 October 1861; *Zanesville Daily Courier*, 26 November 1861; Merrill, *The Soldier of Indiana*, 94; *Zanesville Daily Courier*, 16 October 1861; Nathan Kimball to Joseph J. Reynolds, 4 October 1861, *OR, Series I, Volume V*, 222; *Zanesville City Times*, 17 October 1861.
68. L. M. Shumaker to H. R. Jackson, 4 October 1861, *OR, Series I, Volume V*, 234.
69. *Indianapolis Daily Journal*, 12 October 1861; E. H. C. Cavins to Ann, 4 October 1861, E. H. C. Cavins Letters, Indiana Historical Society; David Poe, *Personal Reminiscences of the Civil War* (Buckhannon, WV: Upshur Republican Print, 1911), 9-10.

70. L. M. Shumaker to H. R. Jackson, 4 October 1861, *OR, Series I, Volume V*, 234; William L. Jackson to Wife, 4 October 1861, William L. Jackson Letters.

71. A. Rust to Henry R. Jackson, 3 October 1861, *OR, Series I, Volume V*, 233; H. R. Jackson to C. R. Stevenson, 7 October 1861, *Ibid.*, 226-27. Although many Confederate participants recorded the results of this action, it is interesting to note that no Union account of the attack on the Confederate left flank could be found.

72. Branner, *Casper Branner of Virginia*, 179; *Indianapolis Daily Journal*, 12 October 1861.

73. Irby Scott to Father, 5 October 1861, Scott Letters; *Richmond Daily Dispatch*, 12 October 1861.

74. Atkins Diary. At least two Confederates, a doctor and his patient, were wounded while within the walls of the hospital. See Hill Diary; *Zanesville Daily Courier*, 17 October 1861; E. H. C. Cavins to Ann, 4 October 1861, Cavins Letters.

75. Thomas Prickett to Matilda Carr, 29 October 1861, Thomas Prickett Papers, Indiana Historical Society; Frank Ingersoll to Sister, 8 October 1861 and 4 October 1861, Ingersoll Letters.

76. *Richmond Enquirer*, 14 October 1861; Recollections of Christian Kuhl; John Henry Cammack, *Personal Recollections* (Huntington, WV: Paragon Printing and Publishing Company, 1923), 39-40; Poe, *Personal Reminiscences*, 10.

77. Cammack, *Personal Recollections*, 38.

78. Milroy Papers, Volume IV, 47; Landon, "The Fourteenth Indiana Regiment on Cheat Mountain," 367-68; *Indianapolis Daily Journal*, 14 October 1861.

79. Samuel V. List Letter, 5 October 1861, List Letters; *Indianapolis Daily Journal*, 21 October 1861; *Cincinnati Daily Commercial*, 9 October 1861.

80. *Staunton Spectator*, 15 October 1861; Atkins Diary, 61; Irby G. Scott to Father, 5 October 1861, Scott Letters; L. M. Shumaker to Henry R. Jackson, 4 October 1861, *OR, Series I, Volume V*, 235; H. R. Jackson to C. L. Stevenson, 7 October 1861, *Ibid.*, 227; Adams, *A Post of Honor*, 78.

81. History of the 32nd Ohio Infantry Regiment, Jacob Pinick Papers, West Virginia and Regional History Collection, West Virginia University.

82. John De Hart Ross Letter, 5 October 1861, John De Hart Ross Letters, Leyburn Library, Washington and Lee University.

83. J. J. Reynolds to L. Thomas, 4 October 1861, *OR, Series I, Volume V*, 227; Frank Ingersoll to Sister, 8 October 1861, Ingersoll Letters; Samuel C. Williams, "General John T. Wilder," *Indiana Magazine of History*, September 1935, 173.

84. *Augusta Chronicle and Sentinel*, 11 October 1861; Dayton, *The Diary of a Confederate Soldier*, 30; Adam W. Kersh to George P. Kersh, 20 October 1861, Kersh Family Letters, *Valley of the Shadow: Two Communities in the American Civil War*, University of

Virginia Library, 3 April 2011 (http://valley.lib.virginia.edu/papers/A0336); Bierce, *Ambrose Bierce's Civil War*, 229. Bierce refers to the soldier as James A. Abbott, who was initially buried on the George Burner farm on the battlefield. His body was later moved to Grafton National Cemetery. Ambrose Bierce, *Battlefields and Ghosts* (Palo Alto, CA: Harvest Press, 1931), 5-6.

85. Author's research utilizing regimental histories, diaries and letters, and newspaper accounts (see Appendices A and B). The returns submitted by the commanding officers were as follows: Federal—eight killed, thirty-five wounded; Confederate—six killed, thirty-three wounded, and thirteen missing. *OR, Series I, Volume V*, 223 and 229 respectively; David Beem Letter, 6 October 1861, David E. Beem Collection, Indiana Historical Society; William Hord Letter, 13 October 1861, William S. Hord Letters, Ohio Historical Society; E. H. C. Cavins to Ann, 4 October 1861, Cavins Letters; Williams, "General John T. Wilder," 173. See also *Indianapolis Daily Journal*, 12 October 1861; Irby G. Scott to Father, 5 October 1861, Scott Letters; Hill Diary.
86. Merrill, *The Soldier of Indiana*, 94; *Cincinnati Daily Times*, 14 October 1861; Harry Milroy Papers, Indiana State Archives; Hill Diary; Irby G. Scott to Father, 5 October 1861, Scott Letters; *Richmond Enquirer*, 15 October 1861.
87. The Battle of Greenbrier River was also known as Camp Bartow, Pine Hill, and Fort Washington.
88. Lucian Barbour Letter, 16 October 1861, Hotchkiss and McCullough Family Papers, Library of Congress; Alf Welton Diary, 3 October 1861, Library of Congress (parentheses in original); *Indianapolis Daily Journal*, 14 October 1861.
89. Milroy Papers, Volume I, 45; *Indianapolis Daily Journal*, 14 October 1861.
90. W. S. Rosecrans to J. J. Reynolds, 17 September 1861, *OR, Series I, Volume V*, 603-04.
91. Culp, *The 25th Ohio*, 30-31.
92. *Augusta Chronicle and Sentinel*, 10 October 1861.

CHAPTER FOUR – A LIVELY MONTH

93. William L. Jackson Letter, 4 October 1861, William L. Jackson Letters; Hill Diary; *Cincinnati Daily Commercial*, 11 October 1861; E. Hannaford, *The Story of a Regiment: A History of the Campaigns, and Associations in the Field, of the Sixth Regiment Ohio Volunteer Infantry* (Cincinnati: The Author, 1868), 138-39.
94. *Augusta Chronicle and Sentinel*, 15 October 1861; Snider Diary; Adams, *A Post of Honor*, 81-82.
95. General Order No. 6, 7 October 1861, National Archives; Joseph J. Reynolds to William S. Rosecrans, 10 October 1861, Records of Department of Western Virginia, Cheat Mountain Division, National Archives.

96. Frank Ingersoll to Sister, October 20, 1861, Ingersoll Letters; *Cincinnati Daily Commercial*, 22 October 1861.
97. *Cincinnati Daily Commercial*, 14 October 1861.
98. Atkins Diary.
99. Snider Diary; Hill Diary.
100. Snider Diary; Hill Diary; *Staunton Spectator*, 15 October 1861.
101. *Richmond Dispatch*, 16 October 1861; Hill Diary.
102. H. R. Jackson to S. Cooper, 12 October 1861, *OR, Series I, Volume V*, 895.
103. F. G. McCoy to Henry R. Jackson, 31 October 1861, H. R. Jackson Papers.
104. F. G. McCoy to Henry R. Jackson, 17 October 1861, H. R. Jackson Papers.
105. Adams, *A Post of Honor*, 97.
106. Snider Diary; Cryder and Miller, *A View From the Ranks*, 11; Hill Diary; *Wheeling Daily Intelligencer*, 4 November 1861; *Cincinnati Daily Gazette*, 26 October 1861; Benjamin Hart Diary, Ohio Historical Society; *Southern Confederacy*, 25 October 1861; Adams, *A Post of Honor*, 98-99; Atkins Diary; *Spirit of Democracy*, 13 November 1861.
107. Hill Diary; *Augusta Daily Chronicle*, 24 October 1861; Dayton, *The Diary of a Confederate Soldier*, 32; Atkins Diary; Snider Diary; Adams, *A Post of Honor*, 101.
108. *Richmond Whig*, 24 October 1861; Hart Diary. Hart recorded that on October 16 the Union troops atop Cheat Summit worked on erecting winter quarters, 20X44 ft. He also noted, "15 of our scouts were out today and were fired on by the cesesh but not any one hurt."
109. *Richmond Whig*, 24 October 1861; Clayton Williams to Ma, 17 October 1861, Lesser Collection.
110. John B. Baldwin to J. P. Benjamin, 16 October 1861, *OR, Series 1, Volume V*, 899-900; Special Orders, No. 140, *OR, Series I, Volume LI, Part II,* 350.
111. Adam Wise Kersh to George P. Kersh, 20 October 1861, Kersh Family Letters, *Valley of the Shadow: Two Communities in the American Civil War*, University of Virginia Library, 3 April 2011 (http://valley.lib.virginia.edu/papers/A0336). Miller had traveled to Staunton in September to procure guns for a planned artillery company. See Adam Wise Kersh to George P. Kersh, 23 September 1861, *Ibid* (http://valley.lib.virginia.edu/papers/A0335).
112. *Cincinnati Daily Commercial*, 5 December 1861.
113. *Wheeling Daily Intelligencer*, 4 November 1861; Hill Diary; Snider Diary.
114. Adam Wise Kersh to George P. Kersh, 20 October 1861, Kersh Family Letters, *Valley of the Shadow: Two Communities in the American Civil War*, University of Virginia Library, 3 April 2011 (http://valley.lib.virginia.edu/papers/A0336); Hill Diary; *Cincinnati Daily Gazette*, 26 October 1861.
115. *Wheeling Daily Intelligencer*, 30 October 1861; Welton Diary; Milroy Papers,

Volume I, 37. In December 1861 Milroy sent Summit to his wife in Rensselaer, Indiana, where he was paid for odd jobs. While in Indiana, Mrs. Milroy taught him how to read and write. Summit desired to serve the Union and in August 1862, he became a servant to officers of the 73rd Indiana Infantry. He eventually volunteered for service in a Colored Regiment and was promoted to sergeant.

116. *Richmond Dispatch*, 30 October 1861; *Indianapolis Daily Journal*, 12 October 1861; General Orders No. 19, 1 November 1861, Edward Willis Papers, The Huntington Library; Samuel V. Fulkerson to Kate, 1 November 1861, Lesser Collection; Michael G. Harman to H. R. Jackson, 20 October 1861, Harman Letters, *Valley of the Shadow: Two Communities in the American Civil War*, University of Virginia Library, 3 April 2011 (http://valley.lib.virginia.edu/papers/A6026).
117. H. R. Jackson to M. G. Harman, *OR, Series I, Volume V*, 912.
118. M. G. Harman to Jefferson Davis, *Ibid.*, 912-13; Joseph A. Waddell, *Annals of Augusta County, Virginia, From 1726-1871* (Bridgewater, VA: Reprinted by C. J. Carrier Company, 1958), 463.
119. *Wheeling Daily Intelligencer*, 10 October 1861; *Richmond Dispatch*, 30 October 1861.
120. William Houghton Letter, 26 October 1861, Houghton Papers; William L. Jackson Letter, 27 October 1861, William L. Jackson Letters; Snider Diary; *Augusta Chronicle and Sentinel*, 2 November 1861.
121. *Indianapolis Daily Journal*, 5 November 1861; *Augusta Chronicle and Sentinel*, 2 November 1861; Snider Diary; *Wheeling Intelligencer*, 1 November 1861.
122. Snider Diary; Adams, *A Post of Honor*, 105.
123. Warner, *Generals in Blue*, 132-33; Welton Diary.
124. *Cincinnati Daily Gazette*, 22 November 1861; *Spirit of Democracy*, 13 November 1861; Welton Diary; Snider Diary. According to Colonel William L. Jackson, three Union soldiers were killed, two were wounded and captured, and eleven muskets were seized. William L. Jackson Letter, 2 November 1861, William L. Jackson Letters. A November 1 letter from a member of the 1st Georgia recorded that one of the Union captives was "a large Irishman, who said that his bloody officers were the cause of his being taken. He said that when he left his camp, he did not know where they were going." The Confederate soldier also noted, "One of the enemy was shot in the head but is still living." Samuel V. Fulkerson to Kate, 1 November 1861, Lesser Collection; *Augusta Daily Constitutionalist*, 10 November 1861.
125. *Wheeling Daily Intelligencer*, 10 November 1861; *Ibid.*, 4 December 1861; Snider Diary; Horace Bell, *Reminiscences of a Ranger: Early Times in Southern California* (Los Angeles, CA: Yarnell, Caystile & Mathes, Printers, 1881).

126. *Wheeling Daily Intelligencer*, 8 November 1861; Frank Ingersoll to Sister, 28 October 1861, Ingersoll Letters; *Cincinnati Daily Gazette*, 22 November 1861; *Ibid.*, 20 November 1861.
127. *Indianapolis Daily Journal*, 12 October 1861; King Journal.
128. *Cincinnati Daily Gazette*, 25 November 1861.
129. *Cincinnati Daily Gazette*, 20 November 1861.
130. King Journal.
131. *Cincinnati Daily Gazette*, 20 November 1861; King Journal.
132. *Ibid.*
133. King Journal; *Indianapolis Daily Sentinel*, 16 November 1861; George Welch Letter, 10 November 1861, George Welch Letters, Indiana State Archives; John Cashner Diary, Indiana Historical Society; Special Order 152, 7 November 1861, National Archives.
134. *Richmond Dispatch*, 22 November 1861.
135. Snider Diary.

CHAPTER FIVE – THE EFFECT OF JACKSON'S PLANNED OFFENSIVE

136. F. G. McCoy to H. R. Jackson, 31 October 1861, H. R. Jackson Papers; Joseph Vanmeter to H. R. Jackson, 1 November 1861, *Ibid.*
137. T. J. Jackson to J. P. Benjamin, 5 November 1861, *OR, Series I, Volume V*, 937.
138. Abstract from return of the Army of the Northwest for the month of October, *OR, Series I, Volume V,* 933; S. Cooper to W. W. Loring, 5 November 1861, *Ibid.*, 938; see also J. P. Benjamin to T. J. Jackson, 6 November 1861, *Ibid.*, 939.
139. Snider Diary; Samuel V. Fulkerson to Kate, 10 November 1861, Lesser Collection.
140. Samuel V. Fulkerson to Henry R. Jackson, 9 November 1861; Edward Johnson Letter, 9 November 1861; James Deshler to Seth Barton, 10 November 1861; Samuel V. Fulkerson to Seth Barton, 10 November 1861, all from H. R. Jackson Papers; Samuel V. Fulkerson to Kate, 10 November 1861, Lesser Collection.
141. Snider Diary; Adams, *A Post of Honor*, 117.
142. *Cincinnati Daily Gazette*, 18 November 1861; Hill Diary.
143. *Cincinnati Daily Gazette*, 22 November 1861; Colonel Robert Driver, ed., "Diary of Capt. J. D. Lilley, Augusta Pioneers, 52nd Regiment Virginia Volunteers," Part II, *Augusta County Historical Society Bulletin. 1991-92.* The soldier captured with Lieutenant Hurt was a Private Jarvis.
144. Snider Diary; *Richmond Dispatch*, 21 November 1861; *Augusta Daily Constitutionalist*, 27 November 1861.
145. Snider Diary; *Zanesville Daily Courier*, 22 November 1861; Atkins Diary.

146. Special Orders, No. 163, Headquarters First Division, Army of Northwest, 18 November 1861, *OR, Series I, Volume LI, Part II,* 382.
147. Snider Diary; Atkins Diary.
148. W. W. Loring to S. Cooper, 19 November 1861, *OR, Series I, Volume V,* 964; S. Cooper to W. W. Loring, 20 November 1861, *Ibid.*, 965.
149. T. J. Jackson to J. P. Benjamin, 20 November 1861, *Ibid.*, 965-66.
150. Irby G. Scott to Father, 26 November 1861, Scott Letters; Atkins Diary, 113; Snider Diary; *Richmond Dispatch*, 30 November 1861.
151. R. H. Chilton to W. W. Loring, 22 November 1861, *OR, Series I, Volume V,* 967. The Union regiments ordered from western Virginia were the 3rd, 6th and 24th Ohio, and the 15th and 17th Indiana Infantry regiments. Unsigned to J. J. Reynolds, 23 November 1861, *Ibid.*, 663.
152. Dayton, *The Diary of a Confederate Soldier*, 36-37; *Augusta Chronicle and Sentinel*, 17 December 1861; Irby G. Scott Letter, 26 November 1861, Scott Letters.
153. Snider Diary.
154. Welton Diary; Thomas Prickett to Matilda Carr, 24 November 1861, Prickett Papers.
155. Irby G. Scott to Father, 26 November 1861, Scott Letters; Snider Diary; *Pittsburgh Gazette*, 11 December 1861.
156. Irby G. Scott to Father, 26 November 1861, Scott Letters; William Batts Letter, 28 November 1861, William Batts Letters, Robert W. Woodruff Library, Emory University; Henry H. Dedrick to Mary E. A. Dedrick, 23 November 1861, Dedrick Letters, *Valley of the Shadow: Two Communities in the American Civil War,* University of Virginia Library, 3 April 2011 (http://valley.lib.virginia.edu/papers/A6003).
157. Betty Hornbeck, *Upshur Brothers of the Blue and the Gray* (Parsons, WV: McClain Printing Company, 1967), 76-77. The children's father, John Yeager, died on 2 December 1861. Many believed he was poisoned. His father, Jacob Yeager, also died during the winter of 1861-62.
158. *Augusta Chronicle and Sentinel*, 17 December 1861; Parson Parker, "Sufferings of the Twelfth Georgia Reg't, in the Mountains of Virginia," West Virginia State Archives.
159. Frank Ingersoll to Sister, 20 November 1861, Ingersoll Letters; Hays, *History of the Thirty-Second Regiment Ohio Veteran Volunteer Infantry*, 183; *History of the 32nd Ohio Infantry Regiment*, Pinick Papers; *Spirit of Democracy*, 27 November 1861.
160. *Spirit of Democracy*, 27 November 1861.
161. VanMeter, *To Water Street With Affection*, 5; Parker, "Sufferings of the Twelfth Georgia."
162. Snider Diary.

163. *Staunton Spectator*, 10 December 1861.

164. J. P. Benjamin to W. W. Loring, 24 November 1861, *OR, Series I, Volume V,* 968-69.

165. W. W. Loring to J. P. Benjamin, 29 November 1861, *OR, Series I, Volume V*, 983-84.

166. Snider Diary; Dayton, *The Diary of a Confederate Soldier*, 40; Atkins Diary; William Batts to Sister, 28 November 1861, Batts Letters.

167. *Baptist Banner*, 20 December 1893.

168. W. W. Loring to J. P. Benjamin, 1 December 1861, *OR, Series I, Volume V*, 975.

169. S. Cooper to W. W. Loring, 5 December 1861, *OR, Series I, Volume V*, 982; J. P. Benjamin to T. J. Jackson, 6 December 1861, *Ibid.*, 982-83.

170. E. Johnson to C. L. Stevenson, 7 December 1861, *OR, Series I, Volume V,* 989-90.

171. W. W. Loring to S. Cooper , 9 December 1861, *OR, Series I, Volume V*, 989; T. J. Jackson to J. P. Benjamin, 9 December 1861, *Ibid.*, 988-89.

172. B. F. Templeton to Brother, 8 December 1861, Benjamin Franklin Templeton Papers, Leyburn Library, Washington and Lee University.

173. Snider Diary; Poe, *Personal Reminiscences*, 11-12; Hart Diary.

174. Snider Diary.

175. Snider Diary; Atkins Diary; Henry H. Dedrick to Mary E. A. Dedrick, 9 December 1861, Dedrick Letters, *Valley of the Shadow: Two Communities in the American Civil War*, University of Virginia Library, 3 April 2011 (http://valley.lib.virginia.edu/papers/A6004); Hart Diary.

176. T. J. Jackson to J. P. Benjamin, 9 December 1861, *OR, Series I, Volume V*, 988-89.

177. General Orders, No. 17, Army of Northwest, 11 December 1861, Willis Papers.

178. Special Orders, No._, Army of Northwest, 13 December 1861, Willis Papers.

179. Schuyler Colfax to George B. McClellan, 12 December 1861, Schuyler Colfax Papers, Indiana Historical Society.

CHAPTER SIX – THE BATTLE OF ALLEGHENY MOUNTAIN

180. The battlefield was also referred to as Camp Allegheny, Camp Baldwin, Black Ridge, and Buffalo Mountain.

181. The battlefield was also referred to as Camp Allegheny, Camp Baldwin, Black Ridge, and Buffalo Mountain. Merrill, *The Soldier of Indiana*, 97; Stevenson, *Indiana's Roll of Honor*, 195.

182. Milroy Papers.

183. Edward Johnson to C. L. Stevenson, 15 December 1861, *OR, Series I Volume V,* 461; Cashner Diary; Hart Diary. Hart recorded the Confederates surrendered on December 7 at Cheat Mountain. Frank Ingersoll to Sister, 15 December 1861, Ingersoll Letters; *Cincinnati Daily Enquirer*, 7 January 1862; Merrill, *The Soldier of Indiana*, 97.

184. Robert Milroy to J. J. Reynolds, n.d. December 1861, *OR, Series I, Volume LI, Part I*, 51.

185. *Zanesville City Times*, 11 January 1862; Stevenson, *Indiana's Roll of Honor*, 196; King Journal.

186. Edward Johnson to C. L. Stevenson, 19 December 1861, *OR, Series I, Volume V*, 462; *Richmond Whig*, 18 December 1861; *Confederate Veteran*, (1913), 21: 26.

187. *Indianapolis Daily Journal*, 27 November 1861.

188. Robert Milroy to J. J. Reynolds, *OR, Series I, Volume LI, Part I*, 51; *Indianapolis Daily Journal,* 25 December 1861; unknown newspaper article, n.d., Julia Gordon Papers, Indiana State Archives; A. W. Kersh to George P. Kersh, n.d., Adam Kersh Letters, Donald Dorr Collection.

189. *Confederate Veteran, Volume XXI* (1913), 26; John G. Keyton Letter, 15 December 1861, John G. Keyton Letters, Perkins Library, Duke University; Snider Diary; Atkins Diary.

190. King Journal; Hamilton, *Recollections of a Cavalryman*, 22.

191. Ambrose Bierce, *The Collected Works of Ambrose Bierce* (New York: Neale Publishing Company, 1912), 232.

192. Moody's official report accompanied Milroy's and the general refers to Moody's reasons without revealing what they were. Unfortunately, the document was apparently lost before the compilation of the *Official Records*, as it is listed as being "Not found." See *OR, Series I, Volume LI, Part I*, 53; Edward Johnson to C. L. Stevenson, 15 December 1861, *OR, Series I, Volume V*, 461; *Pocahontas Times*, 18 July 1935.

193. Hays, *History of the Thirty-Second*, 18; Edward Johnson to C. L. Stevenson, 19 December 1861, *OR, Series I, Volume 5*, 462; Article from unknown newspaper, 16 December 1861, Kirkpatrick Family MSS, 37, Indiana State Archives; Hamilton, *Recollections of a Cavalryman*, 23. A few days before the battle, Private Samuel Pryor of the 32nd Ohio had experienced a premonition of his death. Troubled, he went to Captain Hamilton and asked to pray with him. Hamilton complied but Pryor's premonition proved correct; he was killed early in the battle. See Hamilton, 26.

194. *Pocahontas Times*, 18 July 1935.

195. Edward Johnson to C. L. Stevenson, 19 December 1861, *OR, Series I, Volume V*, 462; G. W. Hansbrough to Edward Johnson, 16 December 1861, *Ibid.*, 465-66; Poe, *Personal Reminiscences*, 12.

196. Robert H. Milroy to J. J. Reynolds, December 1861, *OR, Series I, Volume LI, Part I*, 51; James A. Jones to Robert H. Milroy, 13 December 1861, *OR, Series I, Volume V*, 457-58; Jacob Pinick to John McLaughlin, 3 January 1862, Pinick Papers. Lieutenant Isaiah McDonald was presented with a revolver, holster, and belt

wrested from a Confederate captain during the battle by troops of the 13th Indiana. See *Indianapolis Daily Sentinel*, 25 December 1861.

197. G. W. Hansbrough to Edward Johnson, 16 December 1861, *OR, Series I, Volume V*, 466; Atkins Diary; Hamilton, *Recollections of a Cavalryman*, 24; Poe, *Personal Reminiscences*, 12; Jacob Pinick to John McLaughlin, 3 January 1862, Pinick Papers.
198. Edward Johnson to C. L. Stevenson, *OR, Series I, Volume V*, 462; Snider Diary; G. W. Hansbrough to Edward Johnson, 16 December 1861, *OR, Series I, Volume V*, 466; Poe, *Personal Reminiscences*, 12-13.
199. Snider Diary; William F. Rippetoe to Father, 19 January 1862, Artificial Civil War Collection, West Virginia State Archives; C. D. McCoy to Mother, 14 December 1861, Charles Daniel McCoy Papers, Alderman Library, University of Virginia; *Staunton Spectator*, 31 December 1861.
200. C. E. Dabney to Edward Willis, 15 December 1861, *OR, Series I, Volume V*, 467-68.
201. Snider Diary; Dayton, *The Diary of a Confederate Soldier*, 41-42.
202. *Pocahontas Times*, 18 July 1935.
203. Poe, *Personal Reminiscences*, 13; William Batts to Sis, 16 December 1861, Batts Letters.
204. *Richmond Examiner*, 7 January 1862. Colonel Johnson's club "was composed of a crooked root of a red oak tree and very heavy. It has been sent to the State Library, where it can be inspected by the curious in such matters"; *Pocahontas Times*, 22 April 1909; Edward Johnson to C. L. Stevenson, 19 December 1861, *OR, Series I, Volume V*, 463.
205. Ken and Ilene White to Raymond Watkins, n.d., courtesy of Richard L. Armstrong, Jr.
206. Edward Johnson to C. L. Stevenson, 19 December 1861, *OR, Series I, Volume V*, 462; *Richmond Dispatch*, 18 December 1861; Snider Diary; Robert H. Milroy to J. J. Reynolds, December 1861, *OR, Series I, Volume LI, Part I*, 52.
207. Robert H. Milroy to J. J. Reynolds, December 1861, *OR, Series I, Volume LI, Part I*, 52; King Journal.
208. Snider Diary; William Batts to Sis, 16 December 1861, Batts Letters; Poe, *Personal Reminiscences*, 13; King Journal.
209. King Journal.
210. Snider Diary; Jacob Pinick to John, 3 January 1862, Pinick Papers.
211. *The Land We Love*, May-October, 1867; Recollections of Christian Kuhl; Culp, The 25th Ohio, 5; Hays, *History of the Thirty-Second*, 19-20; Hamilton, *Recollections of a Cavalryman*, 25-26.
212. Snider Diary; Hamilton, *Recollections of a Cavalryman*, 26-27.
213. Stevenson, *Indiana's Roll of Honor*, 198-99.

214. *Richmond Dispatch*, 19 December 1861.

215. Thomas Prickett to Matilda Carr, 1 January 1862, Prickett Papers; *Wheeling Daily Intelligencer*, 24 December 1861; Edward Johnson to C. L. Stevenson, 19 December 1861, *OR, Series I, Volume V*, 463-64; *Indianapolis Daily Sentinel*, 4 January 1862; *Indianapolis Daily Journal*, 25 December 1861; Stevenson, *Indiana's Roll of Honor,* 201-02; Thomas Prickett to Matilda Carr, 23 December 1861, Prickett Papers; *Macon Daily Telegraph*, 4 January 1862.

216. *Indianapolis Daily Sentinel*, 4 January 1862; *Pittsburgh Gazette*, 20 December 1861; Thomas Prickett to Matilda Carr, 1 January 1862, Prickett Papers.

217. Hornbeck, *Upshur Brothers*, 77; *Pocahontas Times*, 18 July 1935.

218. *Pocahontas Times*, 18 April 1904.

219. Robert J. Driver, *1st and 2nd Rockbridge Artillery* (Lynchburg, VA: H. E. Howard, 1987), 101-2; Snider Diary; Adam W. Kersh to George P. Kersh, 16 December 1861, Kersh Family Letters, *Valley of the Shadow: Two Communities in the American Civil War*, University of Virginia Library, 3 April 2011 (http://valley.lib.virginia.edu/papers/A0337).

220. Edward Johnson to C. L. Stevenson, *OR, Series I, Volume V*, 464; John G. Keyton Letter, 15 December 1861, Keyton Letters; *Pocahontas Times*, 18 July 1935. Confederates termed the Union withdrawal as "Slayton's Retreat." Two Confederate soldiers, George B. Sutton and Robert Wolfenbarger, composed a violin tune of this title, in which they claimed they heard John Slayton attempting to order a withdrawal, but he could not verbalize due to stammering.

221. Bierce, *Collected Works*, 233; Poe, *Personal Reminiscences*, 14; Hamilton, *Recollections of a Cavalryman*, 28-30; VanMeter, *To Water Street, With Affection*, 3.

222. *Pocahontas Times*, 3 April 1902; *Richmond Whig*, 18 December 1861. According to the latter article, the 44th Virginia arrived at Allegheny Summit about 6 p.m. W. T. Price Diary, West Virginia State Archives.

223. Cammack, *Personal Recollections*, 43; *Pocahontas Times*, 18 July 1935.

224. *Richmond Whig*, 20 December 1861; A. W. Kersh to George P. Kersh, 25 December 1861, Kersh Family Letters, *Valley of the Shadow: Two Communities in the American Civil War*, University of Virginia Library, 3 April 2011 (http://valley.lib.virginia.edu/papers/A6026); VanMeter, *To Water Street With Affection*, 3.

225. C. D. McCoy to Mother, 14 December 1861, McCoy Papers; VanMeter, *To Water Street, With Affection*, 3; William Batts to Sis, 16 December 1861, Batts Letters.

226. Hornbeck, *Upshur Brothers*, 77; Cammack, *Personal Recollections*, 42; *Staunton Spectator*, 31 December 1861; *Rockingham Register and Virginia Advertiser*, n.d. A member of Wood's battalion, William F. Rippetoe, walked over the battlefield one month after the attack and mentioned in a letter to his father that "the blood is still laying on the log where he [Wood] fell." See William F. Rippetoe

to Father, 19 January 1862, Artificial Civil War Collection, West Virginia State Archives.

227. David Beem Letter, 17 December 1861, Beem Collection; Special Order 211, RG 393 Part 2, 17 December 1861, National Archives. The two officers were Adjutant Patton and 1st Lieutenant Rhodes; *Indianapolis Daily Journal*, 21 December 1861; Julia L. Gordon Papers. Gordon's father later presented a picture of his son to the Pythian Society which was displayed in a hall of a building at Butler College in Indianapolis.
228. Jacob Pinick to John McLaughlin, 3 January 1862, Pinick Papers; author's research utilizing regimental histories, diaries and letters, and newspaper accounts (see Appendices C and D). The returns submitted by the commanding officers were as follows: Federal—twenty killed, one hundred seven wounded, and ten missing; Confederate—twenty killed, ninety-eight wounded, and twenty-eight missing. *OR, Series I, Volume V,* 457 and 468 respectively.
229. *Indianapolis Daily Sentinel*, 4 January 1862.
230. *Zanesville City Times*, 11 January 1862; Frank Ingersoll to Sister, 15 December 1861, Ingersoll Letters; King Journal.
231. *Cincinnati Daily Enquirer*, 7 January 1862; *St. Clairsville Gazette*, 26 December 1861.
232. Robert Milroy to J. J. Reynolds, n.d. December 1861, *OR, Series I, Volume LI, Part I*, 54; Thomas Prickett to Matilda Carr, 23 December 1861, Prickett Papers; *Cincinnati Daily Enquirer*, 7 January 1862; *Zanesville Daily Courier*, 30 December 1861; *Woodsfield Spirit of Democracy*, 1 January 1862; Atkins Diary.
233. *Staunton Spectator*, 24 December 1861.
234. Letter to Alexander H. Stephens and Martin J. Crawford signed by Captains Blandford, Furlow, Hawkins, and Davis, 20 December 1861, Alexander H. Stephens Papers, Robert Woodruff Library, Emory University; Willis A. Hawkins to Alexander H. Stephens, 22 January 1862, *Ibid*.
235. General Orders No__ (no number given), Willis Papers; Dayton, *The Diary of A Confederate Soldier*, 41.
236. C. D. McCoy to Mother, 14 December 1861, McCoy Papers.
237. Edward Johnson to C. L. Stevenson, 13 December 1861, *OR, Series I, Volume V,* 461; Edward Johnson to C. L. Stevenson, 15 December 1861, *Ibid.*, 461-62.
238. A. Stephenson to J. W. Brockenbrough, 15 December 1861, *OR, Series I, Volume LI, Part II*, 414; J. W. Brockenbrough to J. P. Benjamin, 18 December 1861, *Ibid.*, 414.
239. W. W. Loring to Samuel Cooper, 17 December 1861, *OR, Series I, Volume V,* 460.
240. W. W. Loring to Samuel Cooper, 17 December 1861, *OR, Series I, Volume V,* 460; C. L. Stevenson to Edward Johnson, 17 December 1861, *Ibid.*, 1001; Richard

W. Oram, ed., "Harpers Ferry to the Fall of Richmond: Letters of Colonel John De Hart Ross, C.S.A., 1861-1865," *West Virginia History* (1983-84), 45: 169-70.

CHAPTER SEVEN – WINTER WORK

241. John De Hart Ross Letter, 22 December 1861, John De Hart Ross Letters, Leyburn Library, Washington and Lee University; Adam W. Kersh to George P. Kersh, 25 December 1861, Kersh Family Letters, *Valley of the Shadow: Two Communities in the American Civil War,* University of Virginia Library, 3 April 2011 (http://valley.lib.virginia.edu/papers/A6026).
242. Oscar D. Lambert, *West Virginia: Its People and Its Progress* (Charleston, WV: Historical Record Association, 1888), x.
243. *Wheeling Daily Intelligencer*, 3 January 1862.
244. *Zanesville Daily Courier*, 16 January 1862.
245. George Webster to R. H. Milroy, 6 January 1862, *OR, Series I, Volume V*, 497.
246. *Baptist Banner*, 31 January 1894.
247. George Webster to R. H. Milroy, 6 January 1862, *OR, Series I, Volume V*, 497; *Zanesville Daily Courier*, 16 January 1862; *Spirit of Democracy*, 29 January 1862.
248. *Baptist Banner*, 31 January 1894.
249. *Ibid*.
250. George Webster to R. H. Milroy, 6 January 1862, *OR, Series I, Volume V*, 497; *Zanesville Daily Courier*, 16 January 1862.
251. *Spirit of Democracy*, 29 January 1862; *Zanesville Daily Courier*, 16 January 1862.
252. *Spirit of Democracy*, 29 January 1862; George Webster to R. H. Milroy, 6 January 1862, *OR, Series I, Volume V,* 496-98; *Zanesville Daily Courier,* 27 January 1862.
253. *Zanesville Daily Courier*, 16 January 1862; *Spirit of Democracy*, 29 January 1862; *Zanesville Daily Courier*, 27 January 1862; *Staunton Spectator*, 7 January 1862; *Indianapolis Daily Journal*, 10 January 1862. Official reports relating to the Huntersville raid can be found in the *OR, Series I, Volume V*, 496-501.
254. Edward Johnson to Michael G. Harman, 2 January 1862, *OR, Series I, Volume V,* 498-99; Edward Johnson to C. L. Stevenson, 4 January 1862, *Ibid.*, 499-500.
255. Atkins Diary; VanMeter, *To Water Street, With Affection*, 7; *Augusta Daily Constitutionalist*, 14 January 1862.
256. *Staunton Spectator*, 7 January 1862; *Augusta Daily Constitutionalist*, 14 January 1862.
257. *Staunton Spectator*, 7 January 1862; VanMeter, *To Water Street, With Affection*, 5.
258. *OR, Series I, Volume V*, 498; Henry H. Dedrick to Mary E. A. Dedrick, 22 January 1862, Dedrick Letters, *Valley of the Shadow: Two Communities in the American Civil War*, University of Virginia Library, 3 April 2011 (http://valley.lib.virginia.edu/papers/A6006).

259. W. S. Rosecrans to G. L. Hartstuff, 8 January 1862, *OR Series I, Volume LI, Part I,* 517.

260. Jacob Pinick to John, 19 January 1862, Pinick Papers; *Cincinnati Daily Gazette*, 8 January 1862.

261. Frank Reader, *History of the Fifth West Virginia Cavalry, Formerly the Second Virginia Infantry, and of Battery G, First WV Light Artillery* (New Brighton, PA: F.S. Reader, 1890), 52; Don Teter, *Goin' Up Gandy: A History of the Dry Fork Region of Randolph and Tucker Counties West Virginia* (Parsons, WV: McClain Printing Company, 1977), 23; *Wheeling Daily Intelligencer*, 17 January 1862; C. G. Barksdale to George Jackson, 9 January 1862, Willis Papers.

262. From the *Wheeling Daily Intelligencer*, 17 January 1862: "Corporal Wm. Jenkins, slightly in the arm; privates—Frederick Dopp [Patrick Duff], mortally—shot through the left breast, Jas. M. Pfrom, severely—a ball in each leg and one through the left hand, James Whitehair, slight wound in the forehead and a ball through the right arm, John W. Lemon, ball in the leg, and Edward Henderson, shot in the left hand." The paper also printed that a few days following the withdrawal of the Union detachment, Home Guards entered Snyder's home and killed the bedridden Dopp with either bayonets or axes and threw his body into the river. This was discovered to be incorrect. Dopp was captured and taken to prison in Richmond, where he recovered from his near-deadly wound. He was eventually exchanged, and briefly rejoined his company. He was discharged for disability on February 9, 1863. C. G. Barksdale to George Jackson, 9 January 1862, Willis Papers; Homer Floyd Fansler, *History of Tucker County* (Parsons, WV: McClain Printing Company, 1962), 193-94; *Pittsburgh Gazette,* 28 January 1862.

263. Jacob Pinick to Friend, 15 January 1862, Pinick Papers.

264. Jacob Pinick to Friend, 15 January 1862 and 1 February 1862, Pinick Papers; Milroy Papers, Volume I, 10.

265. L. Waddell and John Lewis to John B. Baldwin, 8 January 1862, Willis Papers; Osborne Wilson Diary, Virginia State Archives.

266. Lillian Henderson, *Roster of Confederate Soldiers of Georgia*, 1861-1865 (Haperville, GA: Longino and Porter, Inc., 1960).

267. Miscellaneous Documents, Willis Papers; Henderson, *Roster of Confederate Soldiers of Georgia*, 161. Despite the accusations against Dr. H. K. Green, discharges were given to soldiers of both sides. More than one hundred soldiers of the 12th Georgia were dismissed from service during their time in western Virginia, often due to medical disability, but sometimes because the soldiers were either too old or too young to serve.

268. George W. Hull to C. K. McCoy, 9 January 1862, Willis Papers; Alexander to Edward Johnson, 15 January 1862, *Ibid*.

269. William L. Jackson Letter, 16 January 1862, William L. Jackson Letters; *Baptist Banner*, 20 December 1893; William L. Jackson to Edward Johnson, 18 January 1862, Willis Papers. Jackson stated he would be more valuable with his regiment (31st Virginia), stationed at Camp Allegheny, and recommended that Major A. C. Jones be appointed to command the post at Huntersville, a move which would soon be made.
270. William L. Jackson to Edward Johnson, 18 January 1862, Willis Papers.
271. Edward Johnson to William L. Jackson, 21 January 1862, William L. Jackson Letters. Johnson also stated, "It has not been my intention to keep you there permanently except with a command suitable to your rank, and unless I could find no one suitable, It is much desired by the Citizens that you should remain and in the event of sending a Regiment I might possibly find it necessary to do so."
272. William L. Jackson to Edward Johnson, 25 January 1862, Willis Papers; *Baptist Banner*, 14 February 1894; Edward Johnson to William L. Jackson, 26 January 1862, William L. Jackson Letters. Johnson journeyed to Huntersville during the last week of January. See Adams, *A Post of Honor*, 129. See also Wilson Diary.
273. *Staunton Spectator*, 11 February 1862; *Baptist Banner*, 14 February 1894. The latter article states that a Mr. Alderman was also killed, but other sources indicate his body was discovered on February 9, and that his murder was a separate incident.
274. *Staunton Spectator*, 4 February 1862.
275. Atkins Diary.
276. Henry H. Dedrick to Mary E. A. Dedrick, 22 January 1862, Dedrick Letters, *Valley of the Shadow: Two Communities in the American Civil War*, University of Virginia Library, 3 April 2011 (http://valley.lib.virginia.edu/papers/A6006); Dayton, *The Diary of a Confederate Soldier*, 47; Adam Wise Kersh to George P. Kersh, 29 January 1862, Kersh Family Letters, *Valley of the Shadow: Two Communities in the American Civil War*, University of Virginia Library, 3 April 2011 (http://valley.lib.virginia.edu/papers/A0340); VanMeter, *To Water Street, With Affection*, 3; Atkins Diary; Wilson Diary; Henry H. Dedrick to M. E. A. Dedrick, 8 March 1862, Dedrick Letters, *Valley of the Shadow: Two Communities in the American Civil War,* University of Virginia Library, 3 April 2011 (http://valley.lib.virginia.edu/papers/A6008).
277. Adam Wise Kersh to George P. Kersh, 4 February 1862, Kersh Family Letters, *Valley of the Shadow: Two Communities in the American Civil War,* University of Virginia Library, 3 April 2011 (http://valley.lib.virginia.edu/papers/A0341).
278. Mary E. A. Dedrick to Henry H. Dedrick, 1 February 1862, Dedrick Letters, *Valley of the Shadow: Two Communities in the American Civil War, University of Virginia*

Library, 3 April 2011 (http://valley.lib.virginia.edu/papers/A6019).

279. C. G. Barksdale to George Jackson, 3 February 1862, Willis Papers. Warner, Elborne, and Phares were the three captured Confederates.
280. George Jackson to Edward Willis, 13 February 1862, Willis Papers.
281. Examination of Joseph H. Ebbon taken before Major George Jackson, 15 February 1862, Willis Papers.
282. Joseph H. Lantz to C. G. Barksdale, 3 February 1862, Willis Papers; George Jackson to Edward Willis, 13 February 1862, *Ibid.*
283. George Jackson to Edward Willis, 5 February 1862, *Ibid.*
284. Robert H. Milroy to William S. Rosecrans, 31 January 1862, National Archives.
285. Robert H. Milroy to Wife, February 1862, Milroy Papers. One Union soldier expected to advance toward Staunton "when the roads gets fit." See Jacob Pinick to "Friend John," 20 February 1862, Pinick Papers.
286. Milroy Papers, Volume I, 5-6.
287. *Wheeling Daily Intelligencer*, 24 February 1862; William L. Jackson to Edward Johnson, 11 February 1862, Willis Papers.
288. William L. Jackson to Edward Johnson, 11 February 1862, *Ibid.*
289. William L. Jackson to Edward Johnson, 12 February 1862, *Ibid.*
290. George Jackson to Edward Willis, 17 February 1862, *Ibid.*
291. *Zanesville Daily Courier*, 28 February 1862.
292. Jacob Pinick to William, 22 (addition on 23) February 1862, Pinick Papers; Snider Diary.
293. Adams, *A Post of Honor*, 141; Adam Wise Kersh to George P. Kersh, 1 March 1862, Kersh Family Letters, *Valley of the Shadow: Two Communities in the American Civil War*, University of Virginia Library, 3 April 2011 (http://valley.lib.virginia.edu/papers/A0329)..
294. General Order Number 5, 27 February 1862, Willis Papers.
295. *Staunton Spectator*, 25 February 1862; Dayton, *The Diary of a Confederate Soldier*, 48-49; Snider Diary; Adam Wise Kersh to George P. Kersh, 1 March 1862, Kersh Family Letters, *Valley of the Shadow: Two Communities in the American Civil War,* University of Virginia Library, 3 April 2011 (http://valley.lib.virginia.edu/papers/A0329).
296. R. D. Lilley to William L. Jackson, 28 February 1862, Willis Papers.

CHAPTER EIGHT – CLOSING THE SHOW

297. Reader, *History of the Fifth West Virginia Cavalry*, 154-55.
298. Typescript, Letter, signed "Rusty," Camp Elkwater, March 17, 1862, Hunter Lesser Collection.

299. Jacob Pinick to John, 6 March 1862, Pinick Papers; Harrison Mayberry Calhoun, *Twixt North and South* (Franklin, WV: McCoy Publishing Co., 1974), 16-21; *Wheeling Daily Intelligencer*, 17 March 1862. *In A History of Franklin*, by Elsie Byrd Boggs (Staunton, VA: McClure Printing Company, 1960), the two dead are identified as Bland and Powers; Milroy Papers, Volume I, 17-18; *Wheeling Daily Intelligencer*, 11 March 1862.
300. George Jackson to Edward Johnson, 2 March 1862, Willis Papers; George Jackson to Edward Willis, 3 March 1862, *Ibid.*
301. Snider Diary; Wilson Diary; Cryder and Miller, *A View From the Ranks*, 29.
302. David Fluornoy to B. C. Bouldin, 8 March 1862, Willis Papers.
303. George Jackson to Edward Willis, 10 March 1862; George Jackson to Edward Willis, 11 March 1862; George Jackson to Edward Willis, 16 March 1862, Willis Papers; A. C. Jones to Edward Johnson, 15 March 1862, *Ibid.*
304. George Jackson to Edward Willis, 16 March 1862, *Ibid.*
305. Cryder and Miller, *A View From the Ranks*, 30-31.
306. A. C. Jones to Edward Johnson, 15 March 1862, Willis Papers.
307. A. C. Jones to Edward Johnson, 16 March 1862, *Ibid.*
308. Adams, *A Post of Honor*, 147; General Orders No. 7, 14 March 1862, Willis Papers; Snider Diary.
309. R. E. Lee to Edward Johnson, 15 March 1862, *OR, Series I, Volume V, Part I*, 1101; Edward Willis to R. E. Lee, 18 March 1862, *OR, Series I, Volume XII, Part III,* 828-29.
310. E. Johnson to R. E. Lee, 18 March 1862, *OR, Series I, Volume XII, Part III,* 827-28.
311. R. E. Lee to Edward Johnson, 21 March 1862, *OR, Series I, Volume XII, Part III*, 833-34; William L. Jackson Letter, 14 March 1862, William L. Jackson Letters.
312. W. S. Rosecrans to L. Thomas, 12 March 1862, *OR, Series I, Volume V, Part I,* 744-746; R. H. Milroy to W. S. Rosecrans, 23 March 1862, *OR, Series I, Volume XII, Part III,* 14.
313. R. H. Milroy to W. S. Rosecrans, 16 March 1862, *OR, Series I, Volume XII, Part III*, 3; R. H. Milroy to W. S. Rosecrans, 19 March 1862, *Ibid.*, 7.
314. Robert H. Milroy to William S. Rosecrans, 19 March 1862, *OR, Series I, Volume XII, Part III,* 7; *Wheeling Daily Intelligencer*, 5 April 1862; See also *Wheeling Daily Intelligencer*, 31 March, 1862; *Wheeling Daily Intelligencer*, 11 March 1862.
315. *Wheeling Daily Intelligencer*, 31 March and 5 April 1862; Milroy Papers, Volume I, 24; A. C. Jones to Edward Willis, 25 March 1862, Willis Papers. According to Jones, one of the men was mortally wounded, but he does not note whether it was Albright or Spurgeon.
316. Adams, *A Post of Honor*, 154.
317. A. C. Jones to Edward Willis, 27 March 1862, Willis Papers.

318. A. C. Jones to Edward Willis, 25 March 1862, *Ibid*.

319. Snider Diary.

320. Cryder and Miller, *A View From the Ranks*, 32; Waddell, *Annals of Augusta County*, 467-68; *Baptist Banner*, 7 March 1894.

321. William Skeen to Henry Heth, 4 April 1862 and Henry Heth to John Letcher, 4 April 1862, *OR, Series I, Volume LI, Part II,* 531-32. According to Skeen, Arbogast, Buzzard and Alderman were murdered by two Chewnings and a Cunningham. The Chewnings, particularly Captain Jack Chewning, were already well known due to their involvement in the burning of Sutton on December 29, 1861. A bounty was placed on their heads, and two of the brothers were killed in 1864. Jack Chewning, however, was neither killed nor captured.

322. Cryder and Miller, *A View From the Ranks*, 32-33.

323. R. H. Milroy to John C. Fremont, 31 March 1862, *OR, Series I, Volume XII, Part III*, 34; Cryder and Miller, *A View From the Ranks*, 33.

324. Cryder and Miller, *A View From the Ranks*, 33; R. H. Milroy to John C. Fremont, 7 April 1862, *OR, Series I, Volume XII, Part III,* 57; Milroy Papers, Volume I, 28-29; Hornbeck, *Upshur Brothers*, 78.

325. Milroy Papers, Volume I, 29.

326. Joseph Badger to E. Kelly, 23 April 1865, *OR, Series I, Volume XLVI, Part I,* 1310-14.

BIBLIOGRAPHY
MANUSCRIPTS

Adjutant General's Records, West Virginia State Archives.

James Atkins Diary, Georgia Department of Archives and History.

Lucian Barbour Papers, Library of Congress.

Orion A. Bartholomew Letters, Indiana State Archives.

William Batts Letters, Robert W. Woodruff Library, Emory University.

David E. Beem Collection, Indiana Historical Society.

Lewis Brooks, "Lewis Brooks, Soldier and Citizen," [ca. 1907], Indiana Historical Society.

John Cashner Diary, Indiana Historical Society.

E. H. C. Cavins Papers, Indiana Historical Society.

George Marion Coiner Papers, Virginia Historical Society.

Schuyler Colfax Papers, Indiana Historical Society.

Henry H. Dedrick Civil War Papers, *Valley of the Shadow: Two Communities in the American Civil War,* University of Virginia Library, 15 April 2011. (http://valley.lib.virginia.edu/VoS/personalpapers/documents/augusta/p2dedrickletters.html)

Department of Western Virginia, Records of, Cheat Mountain Division, October-December 1861, and First Brigade, Cheat Mountain Division, Department of West Virginia, Record Group 393 Volume II, and Record Group 94, National Archives.

Samuel V. Fulkerson Letters, W. Hunter Lessser Collection.

Julia Gordon Papers, Indiana State Archives.

William Gray Collection, Virginia Historical Society.

John Levi Griffin Diary, Robert W. Woodruff Library, Emory University.

John Hadley Manuscript, Lilly Library, Indiana University.

John Halvey Letters, Indiana State Archives.

J. L. Harding Letters, Lilly Library, Indiana University.

G. K. Harlow Letters, Virginia Historical Society.

Michael G. Harman Letters, *Valley of the Shadow: Two Communities in the American Civil War*, University of Virginia Library, 15 April 2011. (http://valley.lib.virginia.edu/VoS/personalpapers/collections/augusta/harman.html)

Benjamin Hart Diary, Ohio Historical Society.

John Lyon Hill Diary, Virginia Historical Society.
John William Holloway Papers, Virginia Historical Society.
William S. Hord Letters, Ohio Historical Society.
Hotchkiss and McCullough Family Papers, Library of Congress.
William Houghton Papers, Indiana Historical Society.
Frank Ingersoll Letters, Lilly Library, Indiana University.
H. R. Jackson Papers, Cuyler Telamon Collection, Hargrett Library, University of Georgia.
William L. Jackson Letters, West Virginia and Regional History Collection, West Virginia University.
Kersh Family Letters, *Valley of the Shadow: Two Communities in the American Civil War*, University of Virginia Library, 15 April 2011. (http://valley.lib.virginia.edu/VoS/personalpapers/documents/augusta/p2kershletters.html)
John G. Keyton Letters, Perkins Library, Duke University.
Nathan Kimball Papers, Lilly Library, Indiana University.
James M. King Journal, Indiana Historical Society.
Kirkpatrick Family Papers, Indiana State Archives.
Recollections of Christian Kuhl, Boyd B. Stutler Collection, West Virginia State Archives.
Robert E. Lee Letters, Ron Kolman Collection.
Robert E. Lee Papers, Virginia Historical Society.
Samuel V. List Letters, Indiana Historical Society.
E. J. Lloyd Letter, Artificial Civil War Collection, West Virginia State Archives.
Jacob C. Matheny Papers, Library of Virginia.
Charles Daniel McCoy Papers, Alderman Library, University of Virginia.
Harry Milroy Papers, Indiana State Archives.
Robert H. Milroy Papers, Indiana Historical Society.
Charles S. Morgan Collection, West Virginia State Archives.
Parson Parker, "Sufferings of the Twelfth Georgia Reg't, in the Mountains of Virginia," West Virginia State Archives.
Leonard Perry Papers, Indiana State Archives.
William Pew Letter, Indiana Historical Society.
Jacob Pinick Papers, West Virginia and Regional History Collection, West Virginia University.

W. T. Price Diary, West Virginia State Archives.
Thomas Prickett Papers, Indiana Historical Society.
William F. Rippetoe Letter, Artificial Civil War Collection, West Virginia State Archives.
Eri Rockwell Diary, Indiana Historical Society.
John DeHart Ross Letters, Preston Library, Virginia Military Institute.
John DeHart Ross Letters, Leyburn Library, Washington and Lee University.
Irby G. Scott Letters, Perkins Library, Duke University.
Joseph C. Snider Diary, West Virginia and Regional History Collection, West Virginia University.
Alexander Hamilton Stephens Papers, Robert W. Woodruff Library, Emory University.
Stuart Family Papers, Virginia Historical Society.
A[lmon] Stuart, "The Attack On Fort Washington," Indiana Historical Society.
Benjamin Franklin Templeton Papers, Leyburn Library, Washington and Lee University.
William H. Trisler Letters, Indiana Historical Society.
Augustus M. Van Dyke Papers, Indiana Historical Society.
George Welch Letters, Indiana State Archives.
Alf Welton Diary, Library of Congress.
J. T. Wilder Papers, Indiana State Archives.
Edward Willis Papers, Huntington Library.
Clayton Wilson Letters, Hunter Lesser Collection.
Osborne Wilson Diary, Library of Virginia.
George Washington Wooding Letters, Mary B. Blount Library, Averett College.
E. R. Wyeth Memoranda Book, Indiana Historical Society.

BOOKS

Adams, Charles R. Jr., editor. *A Post of Honor: The Pryor Letters, 1861-1863*. Fort Valley, GA: Garret Publications, Inc., 1989.
Annual Report of the Adjutant General of the State of West Virginia For the Year Ending December 31, 1864. Wheeling, WV: John F. M'Dermot, Public Printer, 1865.

Armstrong, Richard L. *25th Virginia Infantry and 9th Battalion Virginia Infantry*. Lynchburg, VA: H.E. Howard, Inc., 1990.

Ashcraft, John M. *31st Virginia Infantry*. Lynchburg, VA: H. E. Howard, Inc., 1989.

Bell, Horace. *Reminiscences of a Ranger: Early Times in Southern California*. Los Angeles, CA: Yarnell, Caystile & Mathes, Printers, 1881.

Bierce, Ambrose. *Ambrose Bierce's Civil War*. Washington, D. C.: Regnary Gateway, Inc., 1956.

Bierce, Ambrose. *Battlefields and Ghosts*. Palo Alto, CA: Harvest Press, 1931.

Bierce, Ambrose. *The Collected Works of Ambrose Bierce*. New York: Neale Publishing Company, 1912.

Boggs, Elsie Byrd. *A History of Franklin*. Staunton, VA: McClure Printing Company, 1960.

Branner, John Casper. *Casper Branner of Virginia and His Descendants*. Stanford University, CA: Privately Published, 1913.

Calhoun, Harrison Mayberry. *Twixt North and South*. Franklin, WV: McCoy Publishing Co., 1974.

Cammack, John Henry. *Personal Recollections*. Huntington, WV: Paragon Printing and Publishing Co., 1923.

Cochran, Wes. *Pocahontas County WV 1860 Census*. Parkersburg, WV: The Author, 1989.

Cohen, Stan. *The Civil War In West Virginia: A Pictorial History*. Charleston, WV: Pictorial Histories Publishing Co., 1976.

Collier, Calvin. *They'll Do To Tie To! The Story of the Third Regiment Arkansas Infantry, C.S.A.* Little Rock, AR: n.p., 1959.

Cryder, George R. and Stanley R. Miller, compilers. *A View From the Ranks: The Civil War Diaries of Corporal Charles E. Smith*. Delaware, OH: Delaware County Historical Society, Inc., 1999.

Culp, Edward C. *The 25th Ohio Vet. Vol. Infantry in the War for the Union*. Topeka, KS: G. W. Crane and Co., 1885.

Dayton, Ruth Woods, editor. *The Diary of a Confederate Soldier: James E. Hall.* Charleston, WV: Education Foundation, Inc., 1961.

Drake, James Vaulx. *Life of General Robert Hatton*. Nashville, TN: The Author, 1867.

Driver, Robert J. *52nd Virginia Infantry*. Lynchburg, VA: H. E. Howard, Inc., 1986.

Driver, Robert J. *1st and 2nd Rockbridge Artillery.* Lynchburg, VA: H. E. Howard, Inc., 1987.

Fansler, Homer Floyd. *History of Tucker County*. Parsons, WV: McClain Printing Company, 1962.

Farrar, Samuel Clarke. *The Twenty-Second Pennsylvania Cavalry and the Ringgold Battalion, 1861-1865*. Pittsburgh, PA: Published under the auspices of the Twenty-Second Pennsylvania Ringgold Cavalry Association, 1911.

Hamilton, William Douglas. *Recollections of a Cavalryman of the Civil War After Fifty Years, 1861-1865*. Columbus, OH: The F. J. Heer Printing Co., 1915.

Hannaford, E. *The Story of a Regiment: A History of the Campaigns, and Associations in the Field, of the Sixth Regiment Ohio Volunteer Infantry.* Cincinnati, OH: The Author, 1868.

Hays, E. Z. *History of the Thirty-Second Regiment Ohio Veteran Volunteer Infantry.* Columbus, OH: Cott and Evans, 1896.

Henderson, Lillian. *Roster of Confederate Soldiers of Georgia, 1861-1865.* Haperville, GA: Longino and Porter, Inc., 1960.

Hermann, Isaac. *Memoirs of a Veteran Who Served as a Private in the 60's in the War Between the States. Personal Incidents, Experiences and Observations*. Atlanta, GA: Byrd Printing Company, 1911.

Hornbeck, Betty. *Upshur Brothers of the Blue and Gray*. Parsons, WV: McClain Printing Company, 1967.

Keifer, Joseph Warren. *Slavery and Four Years of War*. New York: G. P. Putnam's Sons, 1900.

Lambert, Oscar D. *West Virginia: Its People and Its Progress*. Charleston, WV: Historical Record Association, 1888.

Lang, Theodore F. *Loyal West Virginia from 1861 to 1865*. Baltimore, MD: Deutsch Publishing Company, 1895.

Merrill, Catherine. *The Soldier of Indiana in the War for the Union*. Indianapolis, IN: Merrill and Co., 1864.

Moore, Frank, editor. *The Rebellion Record: A Diary of American Events, with Documents, Narratives, Illustrative Incidents, Poetry, Etc.* New York: G. P. Putnam, 1861.

Moore, George E. *A Banner in the Hills.* New York: Appleton-Century-Crofts, 1963.

Moore, Robert H. *The Charlottesville, Lynchburg and Bedford Artillery.* Lynchburg, VA: H. E. Howard, Inc., 1991.

Moore, Robert H. *The Danville, Eighth Star New Market and Dixie Artillery.* Lynchburg, VA: H. E. Howard, Inc., 1989.

Morton, Oren. *History of Rockbridge County, Virginia*. Baltimore, MD: Regional Publishing Company, 1917.

Pocahontas Historical Society. *History of Pocahontas County, West Virginia*. Marlinton, WV: Historical Society of Pocahontas County, WV, 1981.

Poe, David. *Personal Reminiscences of the Civil War*. Buckhannon, WV: Upshur Republican Print, 1911.

Pool, J. T. *Under Canvas, Or, Recollections of the Fall and Summer Campaign of the 14th Regiment Indiana Volunteers, Col. Nathan Kimball, Western Virginia, in 1861*. Terre Haute, IN: Oliver Barlett, Publisher, 1862.

Price, William T. *Historical Sketches of Pocahontas County, West Virginia*. Marlinton, WV: Price Brothers Publishers, 1901.

Rankin, Thomas. *23rd Virginia Infantry*. Lynchburg, VA: H.E. Howard, Inc., 1985.

Reader, Frank. *History of the Fifth West Virginia Cavalry, Formerly the Second Virginia Infantry, and of Battery G, First WV Light Artillery.* New Brighton, PA: F. S. Reader, 1890.

Robson, John S. *How a One-Legged Rebel Lives: Personal Reminiscences.* Charlottesville, VA: Chronicle Steam Printing Co., 1891.

Smith, Barbara A., editor. *The Civil War Letters of Col. Elijah H. C. Cavins, 14th Indiana.* Owensboro, KY: Cook-McDowell Publications, 1981.

Stevenson, David. *Indiana's Roll of Honor.* Indianapolis, IN: The Author, 1864.

Sturm, Alan. *A Partial History of the Sturm Family of Barbour County, West Virginia.* Buckhannon, WV: n.p., 1973.

Stutler, Boyd. *West Virginia in the Civil War.* Charleston, WV: Education Foundation, Inc., 1963.

Teter, Don. *Goin' Up Gandy: A History of the Dry Fork Region of Randolph and Tucker Counties West Virginia.* Parsons, WV: McClain Printing Company, 1977.

Toney, Marcus B. *The Privations of a Private.* Nashville, TN: The Author, 1905.

VanMeter, John G., editor. *To Water Street, With Affection: The Civil War letters of 2nd Lt. Charles W. Delay, Company E, 25th Virginia Infantry.* Moorefield, WV: The Author, 1999.

Waddell, Joseph A. *Annals of Augusta County, Virginia, From 1726-1871.* Bridgewater, VA: Reprinted by C. J. Carrier Company, 1958.

War Records Office. *War of the Rebellion, A Compilation of the Official Records of the Union and Confederate Armies.* 128 Volumes. Washington, DC, 1880-1901.

Warner, Ezra J. *Generals In Blue: Lives of the Union Commanders.* Baton Rouge, LA: Louisiana State University Press, 1964.

Warner, Ezra J. *Generals In Gray: Lives of the Confederate Commanders.* Baton Rouge, LA: Louisiana State University Press, 1959.

Womack, J. J. *The Civil War Diary of Capt. J. J. Womack.* McMinnville, TN: Womack Printing Company, 1961.

Zinn, Jack. *R. E. Lee's Cheat Mountain Campaign.* Parsons, WV: McClain Printing Company, 1974.

PERIODICALS

Allbritton, Orval E. "The Third Arkansas Regiment From Formation to Fredericksburg." *The Arkansas Historical Quarterly* 16, no. 2 (Summer 1957).

Confederate Veteran.

Driver, Robert J., editor, "Diary of Capt. J. D. Lilley: Augusta Pioneers, 52nd Regiment Virginia Volunteers." *Augusta County Historical Society Bulletin* (1991-92).

The Land We Love.

Landon, William, editor, "The Fourteenth Indiana Regiment on Cheat Mountain: Letters to the Vincennes Sun." *Indiana Magazine of History* 29 (December 1933).

Moore, George, editor, "A Confederate Journal." *West Virginia History* 22, no. 4 (July 1961).

Oram, Richard W., editor, "Harpers Ferry to the Fall of Richmond: Letters of Colonel John De Hart Ross, C.S.A., 1861-1865." *West Virginia History* 45 (1983-84).

Robertson, James I. Jr. "An Indiana Soldier in Love and War: The Civil War Letters of John V. Hadley." *Indiana Magazine of History* 54 (September 1963).

Williams, Samuel C. "General John T. Wilder." *Indiana Magazine of History* 31 (September 1935).

NEWPAPERS

Augusta Daily Chronicle and Sentinel (Georgia)
Augusta Daily Constitutionalist (Georgia)
Barbour Democrat (West Virginia)
Cincinnati Daily Commercial
Cincinnati Daily Enquirer
Cincinnati Daily Gazette
Cincinnati Daily Times

Columbus Daily Enquirer (Georgia)
Indianapolis Daily Journal
Indianapolis Daily Sentinel
Lynchburg Daily Virginian (Virginia)
Macon Daily Telegraph (Georgia)
New York Times
Pittsburgh Post
Pocahontas Times (West Virginia)
Richmond Dispatch
Richmond Enquirer
Richmond Whig
Rockingham Register and Virginia Advertiser
Savannah Daily Morning News
Southern Confederacy (Atlanta, Georgia)
Spirit of Democracy (Woodsfield, Ohio)
Staunton Spectator (Virginia)
Wheeling Daily Intelligencer
Zanesville Daily Courier (Ohio)
Zanesville City Times (Ohio)

APPENDIX A
UNION CASUALTIES-BATTLE OF GREENBRIER RIVER

KILLED: 15

9TH INDIANA

Abbott, Albert I.	Private	C	
Smith, Lewis E.	Sergeant	H	

13TH INDIANA

Hendricks, David J.	Private	K	killed by a shell

14TH INDIANA

Boyd, Amos	Private	C	
Myer, Harrison	Private	H	died of wounds
10/4/1861			
Price, Jonas Urner	1st Sergeant	A*	

**wounded in thigh by shell; leg amputated, died of wounds 10/4/1861*

17TH INDIANA

Dukes, Ezekiel T.	Private	B	cut nearly in two by cannon ball

24TH OHIO

Borgenmeister, Christ.	Private	H	
Kinesel, Jacob	Private	H	
McCann, Nathan	Corporal	B	
O'Neil, John	Private	H	
Riddle, John	Private	B	killed by a six-pound shot

HOWE'S BATTERY

Dougherty, Andrew			lost arm; DOW
Enyart, James			
Price, George L.			

WOUNDED: 35

7TH INDIANA			
Clements, John	Private	B	in the leg
Davidson, John	Private	B	in the arm
Funcell, William H.	Private	C	(possibly Faucett)
Guysinger, Cyrus	Private	H	shot in leg; not serious
James, Alfred	Private	A	
Jones, Thomas	Private	D	slightly
Lanesbury, James	Private	A	
Patterson, Alexander B.	1st Lieutenant	A	slightly
Reynolds, Samuel	Private	B	
West, William	Private	B	slightly
Wooley, William	Private	H	
9TH INDIANA			
Arrick, James	Sergeant	D	
Ball, Thomas L.	Sergeant	F	
Bishop, Henry	Private	G	
Bryant, Isaac S.	Corporal	H	
Clark, Ezra L.	Private	G	
McConnell, Murry	Private	E	
Nafus, John H.	Private	F	
Wilcox, Stephen O.	Private	B	
13TH INDIANA			
Rummell, Jonathan B.	Private	I	slightly
14TH INDIANA			
Foote, L.S.A.	Captain	E	slight wound in arm
Jackson, James S.	Private	D	slightly
Lyon, John D.	Corporal	E	
Smith, Asa	Private	K	

24TH OHIO

Bailey, John	Private	B	slightly
Fuller, William F.	Private	F	slightly
Reiner, Christ.	Private	B	slightly
Sparks			

25TH OHIO

Everingham, John	Private	E	severe ankle wound
Mulgrove, Michael	Private	E	slightly
Pemberton, Elenander	Private	E	slightly

HOWE'S BATTERY

Andrews, Michael F.	Lance Corporal		
Daniels, Cornelius	Private		
Leedridge, John	Private		severely
Rice, George L.	Private		severely

APPENDIX B
CONFEDERATE CASUALTIES-BATTLE OF GREENBRIER RIVER

KILLED: 9

3RD ARKANSAS			
Agnew, John	Private	E	
Amason, Jesse T.	Private	E	
Blocker, William O.	Private	H	
1ST GEORGIA			
Brown, David W.	Private	F	
12TH GEORGIA			
Crye, Richard	Private	K	
Terrell, D.W.	Private	I	died of wounds
31ST VIRGINIA			
Munford, Jack	Private	A	died from wounds
RICE'S BATTERY			
Graves, Benjamin Franklin	Sergeant		
Tidler, Samuel Luther	Private		

WOUNDED: 34

3RD ARKANSAS			
Calhoun, Milton	Private	A	slightly
Scott, Anderson C.	Private	I	mortally
Stephens, S.T.	Private	A	slightly
1ST GEORGIA			
Beck, David L.	Private	H	severely in left groin; died in 1861

12TH GEORGIA

Anders, Wyatt F.	Private	K	slightly
Browning, Rowan J.	Private	A	
Dean, John T.	Private	E	three flesh wounds
Ware, D.W.	Private	E	slightly

23RD VIRGINIA

Fleming, William	Private	K	severely, burst shell
Reinhart, Gustavus	Private	H	severely in right leg

31ST VIRGINIA

Bartlett, Frederick W.	Lieutenant	A	severely
Exline, Laban Radcliff	Corporal	A	slightly
Keller, Isaac	Private	D	missing, wounded
Mathew, Eugenius	Private	K	missing, wounded
Perry, John O.	Sergeant	B 2	
Phillips, John Riley	Lieutenant	K	
Shaver, Benjamin B.	Corporal	A	severely
Slaton, William W.	Private	G	severely, missing

44TH VIRGINIA

Fitzgerald, Stephen	Private	F	slightly
Kane, (Kaid?)	Private		slightly
Kelly, John	Private	E	slightly
Mann, Edward N.	2nd Lieutenant	H	slightly
Payne, John H.	Private	E	slightly

CHURCHVILLE CAVALRY

Parsons, Nelson	Private		

RICE'S BATTERY

Crank,			
Rice, William H.	Captain		foot amputated
Robey, Thomas			wounded in hand
Stillwell, George A. or William			slightly

SHUMAKER'S BATTERY

Dickenson, Joseph R.			severely; shell in thigh
Earles, Alexander M.			
Eaton, Calvin H.P.	Corporal		severely in right thigh
Elliott, Thomas A.			piece of shell in thigh
Shumaker, Lindsey Mayo	Captain		slightly
Wimsey, Thomas G.			slightly; minie ball in thigh

CAPTURED: 16

3RD ARKANSAS

Brooks, James W.	Private	C	captured (wounded?)*
Carter, John C.	Private	H	captured*
Garron (Gaston), John	Sergeant	C	captured*
Harris, J.S. (Isaw?)	Private	D	captured*

31ST VIRGINIA

Alford, James H.	Private	K	captured*
Alford, Thomas	Private	K	POW
Arbogast, Jefferson	Private	F	captured
Evans, Evan	Private	A	captured*
Gainer, Solomon	Private	K	prisoner*
Hoffman, A. Ely	1st Sergeant	A	captured*
Morgan, George P.	Private	A	captured*
Nay, James Henson	Private	A	MIA, POW*
Radosky, James	Private	I	captured
Thompson, Josiah	Private	H	captured*
West, Thomas	Private	A	captured
Wolf, Phillip	Private	G	captured*

**On "List of Prisoners Captured," National Archives.*

APPENDIX C
UNION CASUALTIES-BATTLE OF ALLEGHENY MOUNTAIN

KILLED: 30

9TH INDIANA

Gordon, Joseph	Private	G
Huntingden, Benjamin F.	Corporal	H
Kilmer, Jackson	Private	A
Mackey, Thomas R.	Sergeant	E
Nouls, Perry	Private	G
Pangborn, Walter H.	Private	F
Souder, David S.	Private	A
Wilson, Charles	Private	G

13TH INDIANA

Bogne, Thomas	Private	E
Bower, Jacob	Private	I
Day, William	Private	D
Fogen, Mathew	Private	B
Jones, Joseph P.	2nd Lieutenant	K
Stinson, William	Private	F
Wessa, Jacob	Private	I

25TH OHIO

Briscoe, John H.	Private	K
Decker, Marcus L.	Corporal	K
Fuller, John C.	Private	F
Hayden, Jonathan	Private	I
Latham, Charles	Private	D
Maher, William S.	Private	G
Nye, Isaac	Private	E
Richards, John	Private	A
Shacklee, Lorenzo	Private	I
Stewart, Levi S.	Corporal	E
Thayer, Christopher J.	Private	E

32ND OHIO

Clark, William	Private	F	
Pryor, Samuel H.	Private	G	

2ND (WEST) VIRGINIA

Sickman, Alfred	1st Lieutenant	G	
Steider, Gustavus	Private	A	

DIED OF WOUNDS: 13

9TH INDIANA

Allen, Charles H.	Private	E	02/13/62
Widener, David	Private	K	01/05/62

13TH INDIANA

Burns, John	Private	E	06/06/62
Shirley, William	Private	E	02/19/62

25TH OHIO

Beckette,Elijah	Private	C	01/29/62
Henthorn, William	Sergeant	C	12/18/61
McMullen, James	Private	A	01/10/62
Ogden, Gilbert J.	Private	G	08/23/62
Page, Harlem	Private	K	02/27/62

32ND OHIO

Hamilton, Robert J.	Private	G	12/28/61
Lipsy, Abraham	Private	F	02/14/62

2ND (WEST) VIRGINIA

Heist, John L.	4th Corporal	E	
Steward, James	1st Corporal	F	01/10/62

WOUNDED: 97

9TH INDIANA

Bliss, Christopher	Private	I	
Dewire, Benjamin	Private	K	
Heckerthorn, William	Private	I	
Hill, Elijah B.	Private	H	
Madden, Thomas	Captain	A	
Moore, Alron P.	Private	G	
Packard, Leverene	Private	I	
Richards, Moses E.	Private	D	
Rusk, Frank M.	Sergeant	K	
Sanders, Erastus	Private	C	
Sherwood, James R.	Corporal	E	
Sweet, William	Private	B	
Wilcox, Stephen	Private	B	

13TH INDIANA

Cole, Jesse B.	Sergeant	E	shot through both arms
Durst, William A.	Sergeant	H	slightly
Fisher, Ferdinand	Private	I	slightly
Foster, Edward	Sergeant	C	seriously
Hines, George W.	Private	E	left hand, right breast
Howbert, Malantheon	Private	E	flesh wound in left thigh
Kath, John	Private	K	slightly
Kemp, David	4th Sergeant	E	shot in side of neck
Kenyon, Benj.	Corporal	I	slightly
Lane, Stephen	Private	E	slightly on left hip
Long, Columbus	Private	E	slightly
Long, Lewis	Private	E	shot in the back
Miller, James K.	Private	A	slightly
Mitchell, Elijah R.	Private	G	slightly
Nelroman, John	Private	K	seriously
O'Neal, William	2nd Lieutenant	H	shot in thigh
Ring, George L. J.	3rd Sergeant	E	flesh wound in left leg
Spencer, Dennis	Private	G	slightly
Swindler, Harry H.	Corporal	I	slightly

25TH OHIO

Aulter, George M.	Private	F	slightly in elbow
Barlow, William C.	Private	I	slightly
Barrett, Cornelius S.	Corporal	H	severely in face
Beck, Charles	Private	B	left arm fractured
Bolen, James C.	Private	A	slightly
Breach, James	Private	I	slightly in leg
Brown, William H.	Private	D	
Butler, Levi	Private	A	slightly
Chadwick, William	Private	H	elbow shattered
Coe, Daniel S.	Corporal	D	severely
Craig, David	Private	H	
Crooks, Daniel J.	Corporal	A	slightly
Cumpton, Robert B.	Private	D	slightly
Dirlam, Derius	1st Lieutenant	D	slightly
Dunn, Jonathan	Private	C	supposed mortally
Durson, John P.	Private	H	severely in arm
Ewalt, John	Private	G	slightly in right arm
Fisher, John D.	Private	G	slightly on top of head
Freeh, August	Private	E	slightly on head
Haney, George	Private	G	supposed mortally
Hapton, Joseph J.	Private	B	slightly
Harris, Michael	Private	G	slightly in neck
Henry, Samuel	Private	A	slightly
Henthorn, W. J.	Private	C	slightly
Hesson, Enville A.	Corporal	F	slightly in elbow
Holland, John W.	Private	A	seriously
Hutchinson, Andrew	Private	K	slightly
Jones, Benjamin F.	Private	D	supposed mortally
Jones, Thomas	Private	F	thigh fractured
King, Charles H.	Private	A	seriously
Kinkead, Blair	Private	H	severely in calf of leg
Kirk, Isaac		J	slightly
Lewis, Sheppard	Private	K	supposed mortally
Lingo, Archelus	Private	J	slightly in hand
Lockwood, William T.	Private	A	slightly

Lovett, Nelson C.	Private	I	slightly
Martin, George W.	1st Sergeant	B	slightly
Masters, Isaiah	Private	C	slightly
McKinley, John	Private	F	slightly in hand
Meek, Henry	Private	A	slightly
Meredith, Asa	Private	F	severely in shoulder
Merryman, John D.	2nd Lieutenant	B	slightly
Phelps, Richard D.	Private	E	severely in thigh
Read, George W.	Private	H	severely in cheek
Rearick, John E.	Private	E	severely in leg
Rodier, Charles C.	Private	D	
Ryan, Levi	Private	A	slightly
Thomas, Hezekiah	Sergeant	A	seriously
Ward, Hiram A.	Sergeant	D	supposed mortally
Ward, John B.	Corporal	D	severely
White, William	Private	D	slightly
Wilson, Hugh	Sergeant	F	toe shot off

32ND OHIO

Brant, Charles C.	1st Lieutenant	F	slightly
Brosens, Harper	Private	G	seriously
Clarke, John	Private	F	slightly
Hamilton, Isaac	Private	B	slightly
Hess, Thomas B.	Private	K	seriously*
Pryor, Chas.	Private	H	seriously
Sharpe, Will	Private	F	slightly
White, James	Private	G	slightly

**Hess died some years thereafter from effects; comrades carried him from Allegheny to Cheat Mountain on a crude litter, where he remained until able to be taken home.*

2ND (WEST) VIRGINIA

Hulville, William	Private	G	wounded in cheek
Mayson, M.B.	Private	D	shot through body
Reynolds, Samuel L.	Private	A	wounded in shoulder

Schweder, Christian	4th Sergeant	C	dangerously
West, Oliver R.	2nd Lieutenant	A	shot in the knee

MISSING: 2

13TH INDIANA

Brown, William	Private	D	captured
Randolph, Lott	Orderly Sgnt.	E	missing

APPENDIX D
CONFEDERATE CASUALTIES-BATTLE OF ALLEGHENY MOUNTAIN

KILLED: 20

12TH GEORGIA

Carter, Abner W.	Private	A
Davis, William H.	Private	G
Dyer, W. Scott	Private	H
Hadaway, Wiley D.	Private	K
Moore, Henry E.	1st Lieutenant	I
Phelps, Augustus	Private	K

31ST VIRGINIA

Nichols, Henry C.	Private	A
Nutter, John Alvin	1st Sergeant	C
Smith, Ethelbert	Corporal	C
Smith, James L.	Private	C
Whitman, George W.	Private	C

HANSBROUGH'S AND REGER'S BATTALIONS

Green, John C.	Private	C (H)
Hefner, Augustine	Private	B
Mollohan, William H.	Captain	G2 (H)
Pullens, James W.	Private	F2
Reger, Jerome J.	1st Lieutenant	B
Tinney, Thurman	Private	C
Wood, John T.	Private	D2 (R)

ARTILLERY

Anderson, Pierce B.	Captain	Lynchburg Artillery
McKown (or McCown), J.H.	Private	2nd Rockbridge Artillery

DIED OF WOUNDS 14

12TH GEORGIA

Little, A.F.	2nd Sergeant	G	
Middleton, Alexander	Private	G	

31ST VIRGINIA

Cropp, Aldridge J.	Corporal	C	bowels, thigh, shoulder
Cross, Daniel	Private	K	
Sheets, Isaac	Private	G	
Springstone, Henry V.	Private	D	
Tennant, Lemon	Private	A	
Thompson, Lewis Stemrod	Lieutenant	A	

52ND VIRGINIA

Baskins, William S.	wounded in hip; died 2/13/62
Welch, Henry D.	wounded in leg while on picket 12/12/61; leg amputated; died 2/13/62

HANSBROUGH'S (H) AND REGER'S BATTALIONS

Collins, Martin V.B.	Private	A(H)	shot in head
Dyer, Zebulon	Sergeant	H	
Young, John B.	Private	C	

2ND ROCKBRIDGE ARTILLERY

Griffin, Lewis A.	Private	leg amputated; died of wounds

WOUNDED: 93

Johnson, Edward	Colonel	slightly wounded
Deshler, James	Captain	shot through both thighs

12TH GEORGIA

Arnold, William T.	Private	G	
Batts, William	Private	A	slightly in calf
Burnett, Christopher C.	Private	I	wounded in eye; lost sight

Carson, John James	Private	D	lost finger
Corley, William G.	Private	I	
Crittenden, Thomas J.	1st Sergeant	B	
Daniel, J.A.	Private	A	wounded in eye; lost sight
Davis, Edward S.	4th Sergeant	G	
Glover, John T.	Private	B	
Green, H.K.	Surgeon		slightly wounded in hand
Hall, Bolin G.	Private	I	leg amputated below knee
Harvey, Moses J.	Corporal	K	
Haskins, James Woodson	Private	G	
Lainey, Noah S.	Private	D	bruised on neck
Marshall, Asa Monroe	Private	G	
McKay, Henry Kent	2nd Lieutenant	A	
McLeroy, David D.	Private	G	
Ogletree, Absalom H.	Private	A	
Pickett, Robert L.	Private	A	
Rodan, William	Private	H	
Smith, James A.	Private	I	severely
Taylor, Jesse M.	Private	A	
Thacker, James A.	Private	K	
Thomas, Lovic H. (Tuck)	Private	H	
Thompson, Wiley Y.	2nd Lieutenant	F	
Wallace, Nathaniel H.	Private	A	
Worrell, Joseph W.	Private	K	wounded in leg; amputated
Zachry, Abner Rowan	1st Corporal	G	

31ST VIRGINIA

Bagby, Geo. A.	Private	I	or sergeant major
Beveridge, George W.	Private	G	
Bird, John W.	Corporal	E	
Bond, Osterman	Private	C	
Bruffy, Patrick W.	Private	I	or lieutenant

Compton, Thomas A.	Sergeant	I	
Dawson, Martin Luther	Private	C	
Galvin, James B.	Lieutenant	I	
Golden, M.	Private	H	
Hall, David H.	Private	I	
Harris, Wm.	Private	I	
Helmick, A.	Private	B	
Houghton, James W.	Private	I	(or Howton)
Huggins, Samuel C.	Private	G	
Johnson, Isaac	2nd Lieutenant	H	
Kerr, James S.	Private	A	
Lake, Granville C.	Private	C	
Lockridge, Andrew Jackson	Private	B2	severely in shoulder
McLaughlin, Robt. W.	Private	D	slightly
McNemar, Washington Byron	1st Lieutenant	I	
Moore, Martin	Private	K	
Mundel, Frank	Private	A	
Phillips, John Riley	Lieutenant	K	
Powers, Newton J.	Private	I	
Pridemore, James W.	Private	C	Blacksmith
Simms, Alfred	Private	C	seriously through pelvis
Smith, Nathan S.	Private	I	
Snider, Joseph C.	Private	C	wounded 3 times buckshot
Taylor, Wm. S.	Private	C	
Toothman, Davis	Lieutenant	A	
Tucker, Jacob	Private	A	
Tucker, Samuel J.	Private	A	
Wilfong, Elisha	Private	G	

52ND VIRGINIA

Bowers, David	Private	wounded in leg 12/12/61

HANSBROUGH'S (H) AND REGER'S (R) BATTALIONS

Boggs, Edwin (or Edward)	Sergeant	E2(R) lost left arm

Bragg, Irvin (or Irwin) W.	Private	G2(H)	left leg amputated
Brown, Isaac N.	Private	G2(H)	
Camden, Edwin Duncan	Captain	C2(H)	spent ball in elbow
Dyer, Charles Edmond	3rd Lieutenant	E (R)	wounded slightly
Halderman, John W.	Private	A2(H)	wounded 3 or 4 times
Hamilton, David	Private	D2(R)	wounded slightly
Hansbrough, George W.	Lieutenant Colonel		shot by pistol
Haymond, Luther D.	2nd Lieutenant	G2(H)	arm badly shattered
Hedrick, Henry J.	Private	E (R)	wounded slightly
Hillery, Regnal (or Riglin)	Private	B2(R)	slightly in hand
Lowman, William Henry	1st Corporal	H2(R)	
Lucas, Peter F.	Private	H2(R)	wounded slightly
McKemy, Robert Alfred	1st Sergeant	H2(R)	wounded slightly
Mitchell, Abel	Private	E (R)	wounded slightly
Moneymaker, Archibald B.	Private	H2(R)	
Murphy, Herbert	Private	A2(H)	
Musgrove, Pinkney A.	Private	A2(H)	
Nicely, Morton	Private	H2(R)	lost thumb, three fingers
Nicely, Salome	Private	H2(R)	lost foot
Poe, David	1st Lieutenant	A2(H)	wounded in right leg
Rader, William M.	3rd Sergeant	G2(H)	
Snider, John Jack	Private	H2(R)	wounded slightly
Young, William B.	Private	C2(H)	

PITTSYLVANIA CAVALRY

Nuckols, John	Private		slightly in both hands and left arm

2ND ROCKBRIDGE ARTILLERY

Lusk, John A. M.	Lieutenant		
Sloan, John Cooke	Private		
Strickler, Archibald W.	Sergeant		
Wallace, Albright	3rd Corporal		left arm

CAPTURED OR MISSING IN ACTION: 34

12TH GEORGIA

Lanier, William	Sergeant	K	captured
Torbert, William L.	Private	A	captured
Ward, Uriah G.	Private	B	captured

31ST VIRGINIA

Arbogast, Jacob	Private	B	prisoner
Ashby, James	Sergeant		captured
Benson, William W.	Sergeant	B2	captured
Bird, Otho M.	Private	E	captured
Campbell, David H.	Private	E	prisoner
Chew, Jonas Watts	Private	B2	captured
Connelly, Francis M.	Private	D	captured
Golden, James Madison	Private	H	captured
James, Ashbel F.	Sergeant	C	captured
Rexrode, Solomon Jr.	Private	E	captured
Talbott, Perry M.	Private	H	captured
Thompson, Amos G.	Private	B	captured
Thompson, George W.	Private		captured

52ND VIRGINIA

Able, Mathias	Private	G	captured while on picket
Carter, Charles B.	Private	B2	captured while on picket
Coyner, George Adam	2nd Corporal	B2	captured while on picket
Moses, Samuel M.	Private	B2	captured
Shifflett, Pemberton C.	Private	B2	captured
Sillings, Armstrong R.	Private	B2	captured
Ward, Levi	Private		captured

HANSBROUGH'S (H) AND REGER'S (R) BATTALIONS

Beverage, Thomas	Private	F2	missing in action

Blankenship, James J.	Private	C2(H) captured
Bragg, Edward S.	Private	A2(H) missing in action
Cutlip, F. Leonard	Private	C2(H) captured
Hall, James H.	Private	C2(H) captured
McCutchan, John Franklin	3rd Corporal	H2(R) captured
McQuain, Hugh	Private	H2(R) captured
Morris, George S.	Private	A2(H) missing
Paugh (or Pugh), William	3rd Sergeant	C2(H) captured
Rexroad, Laban	Private	K2 captured
Ware, Levi S.	Private	C2(H) captured

APPENDIX E
POETRY

1. "The Attack On Ft. Washington, Oct. 3, 1861."

Written by Private Almon Stuart, Company I, 9th Indiana Infantry
Indiana Historical Society

Come all you gallant Hoosier crew,
And listen to my rhymes.
While I a song now sing to you,
Relating to the times,
Of how in old Virginia,
Hard fighting was our sport,
And how the rebels ran away
For shelter in their fort.
Oh! Run poor rebels, hurry up,
The Ninth is at your back,
And Milroy's not the man to stop,
When once upon the track.

'Twas on a dark October night,
The Second I believe;
When with a shout of wild delight,
We our orders did receive
To march, to fight the rebel foe,
Who were several thousand strong;
And who in fortress lay, twelve miles away,
In a fort called Washington.
Poor rebels, each one hunt your den;
The Fourteenth onward rolls,
With Kimball and their [sic] the men,
To make you hunt your holes.

Onward o'er mountains and through mud,
Our gallant men march on,
Resolved to wade through streams of blood,

Ere they would ever run;
Seven miles away, 'tis break of day,
The Ohio Thirty second have
With two brass guns to the right away,
March toward the rebel's rear.
Poor rebels now you'd best look out,
The Thirty-second comes,
With Ford, to ferret traitors out,
And send them to their tombs.

Now down a long and winding hill,
Our forces take their way,
H of the Ninth are scouts; they will
Most surely clear the way;
To Green Brier river they have come,
To the bridge across the stream,
Here many a rebel finds a tomb,
Where Northern rifles gleam.
Poor rebels you're too far from fort,
The Fifteenth is at hand,
With Wagner, and with them 'tis sport
To sweep you from the land.

Hark! 'tis the thunder of our guns,
And see the streams of fire
Pour from our ranks, their picket runs,
The rebel crew retires.
With shot and shout we still rush on,
Now on a hill we form,
While Howe's and Loomis' big guns,
Send forth an iron storm.
Oh! Rebels seek a shelter now,
The Seventh comes thundering on,
With Dumont, and they just know how
To make you rebels run.

Across Green Brier's pleasant vale
Our forces now deploy,
Ditches we leap and fences scale,
And loudly shouts each boy;
While all the time an iron storm
Is flying over head
And many a true and gallant form
Is numbered with the dead.
Oh! Rebels we are getting nigh,
The Thirteenth you may see,
With Sullivan, and they'll stand by,
The banner of the free.

Now on the plain behind a fence,
The Ninth and Fourteenth lay,
While our batteries scarce two rods hence
Are thundering away;
The foe directly in our front,
Not one mile off reply,
Some of them safe shelter hunt,
And some in terror fly.
Poor rebels, woful is your plight,
The Seventeenth is about,
With Hascall, and they're sure to fight
As you will soon find out.

Their guns have almost ceased their fire,
The Fort is almost won,
To the hillside now our men retire,
Soon will the fight be done;
Hark! 'tis a shout comes bursting out,
From rebel throats we hear;
Aid on to the foe has come no doubt,
All now is lost we fear.
Yet rebels keep behind your works,
Ohio's Twenty-fourth is nigh,

With Ammond, and they fight like Turks
And don't know how to fly.

Our guns have ceased our powder gone,
And slowly we retire;
Six hours hard fighting we have done,
Six hours of constant fire;
But yet our loss is very small,
How many I can't say,
Less than a score, I think in all,
And those we brought away.
Poor Rebels follow if you dare,
The Ohio Twenty-fifth is round,
With Jones, and rebels they'll not spare,
Wherever they are found.

2. A member of Company B of the 24th Ohio penned the following verses in memory of Nathan McCann and John Riddle, who were killed during the Battle of Greenbrier River.
Zanesville Daily Courier, 16 October 1861

Here's to the memory of our boys,
That long were by our side
That marched bravely to the field,
On which they nobly died.

The good "old flag" for which they fought
Above them was unfurled
That flag which has for many years
Been honored by the world.

They longed to fight for those dear rights
The rights all should maintain;
That right of making faultless laws,
And get no armed blame

For those dear rights they, fighting, fell.
And murmured not a word.
"My God have mercy on our souls,"
Was most of what we heard.

They soon were freed from all their pains
Their spirits soon died out;
For God did take them unto him,
To be forever blest.

Dear friends you must not weep for them;
Though near to you they were.
They died a true and glorious death,
Their county's foes to dare.

On fame's long roll their names shall
In ages yet to come,
And "children's children yet unborn"
Shall thank them for their freeland home.

3. "Lines on the Death of Serg't Price, and Respectfully Inscribed to Co. A, 14th Reg't Ind. V. M."

Written by W. H. A.
Cheat Mountain, Va., Oct. 6, 1861
J. T. Pool, Under Canvass (Terre Haute: Oliver Bartlett, 1862), 49-50

O! heard you that music sound sad on the hill,
Where a band cometh solemnly, slowly and still,
'Tis Company A, of the gallant Fourteenth.
And they bury their comrade, the soil underneath.

He hath fallen in battle, his feet to the foe,
His life blood it crimsons the green moss below,
But he grasps his good musket, still eager for fight,
And to battle for Freedom and Union and Right.

But the death dealing ball hath shattered his life,
And his comrades will bear him from scenics filled with strife,
While a prayer from his heart ascends up to Heaven,
That our Union by traitors, may never be riven.

He hears the last shot that is fired by the foe,
He sees the retreat quiet, steady and slow,
And the prayer from his heart still ascends to the Throne,
That the might of the right in Thee may be shown.

They have now laid him down on the bed of his death,
While his Captain bends over him to catch his last breath,
And bear to his friends in our noble old State,
The last words of him who hath met this sad fate.

Tell them he murmured, his breath coming fast.
I battled for Freedom and Right to the last,
I bear to my friends the name that I gained,
On the field where the foe their shot and shell rained.

Tell them God from His Throne in the sky,
Looks down on us here with His "All Seeing Eye."
And in His good time will aid in our cause,
For Union and Freedom, Constitution and Laws.

He said and fell back; and as death glazed his eye
His soul it ascended to glory on high,
Where it stands by the Throne and implores the great God
To smite those, who fight 'gainst the right, with his rod.

On the soil of Virginia they laid him to rest,
Where the rude winds of Winter will sweep o'er his breast,
And his comrades will think as deep sighs their breasts rend,
Of the soldier, their brother and brave and true friend.

The night watch will pace past his rude lowly grave

And think how he died his dear country to save,
And his heart will in silence a firm resolve form
To fight till our Union is freed from the storm.

4. Nearly twenty years after his death, Sarah T. Bolton Reese wrote a touching poem to the memory of Joseph R. T. (Josie) Gordon, who was killed at the battle of Allegheny Mountain.
Julia Gordon Papers, Indiana State Archives

From the wall of a stately mansion,
A fair young face looks down,
Bedight with the cap of a soldier,
That seemeth a fitting crown,
With firm resolve on his beardless lip,
And truthful eyes dark brown.

By the smouldering fire of genius,
By the manly strength and grace;
By the innate truth and honor
Painted on his boyish face,
You know that his blood claims kindred
With the blood of a noble race,

That his mind and heart are dowered,
With a wealth beyond fine gold,
That came from the souls of heroes
Warm, generous true and bold,
Who fought for the right and conquered
In the shadowy days of old.

There needeth no gift prophetic
No wave of Sorcery's wand
To show that his future pathway
Will be upward, onward and grand
To the loftiest plane of manhood,
Where the good and true may stand.

He marched away, with the bravest,
To the battle's seething hell,
Unawed by the cannon's thunder,
Or the hail of shot and shell;
And there in the fiery tempest,
The fair boy fought and fell.

Would you know whose life-like semblance
Is shrined in that golden frame?
On his country's Roll of Honor,
Ye will find inscribed the name
Of a hero boy, who bravely died
Defending his country's fame.

Who marched to the field of battle
When the sunshine of life's way
Unclosed the hearts of the roses
That bloomed along his way,
And the beautiful Hope of boyhood
Dreamed dreams of a long bright day.

To his father's tender letter,
Half regretting he had gone
To the hardships and the dangers,
Shunned by many an older one,
The boy's reply was worthy,
Of the noblest Roman's son.

Thus it ____ Remember, father,
That your counsel sowed the seed;
In the training of my childhood
____ ripened to this deed,
In behalf of our dear Union,
In its time of utmost need.

"You have taught me to love my country,

Trained me to be brave and true,
In every word and action,
That a man may say or do,
For the sake of human freedom.
Father, I but follow you."

But, alas! For the brave young spirit,
For its hope and high emprise,
For the hand that has lost its cunning
For the light of the dark brown eyes.
For the promise of noble manhood;
Alas, for the sacrifice!

We turn from the life like picture
With a sense of sorrow deep,
From the still, invisible presence
We turn away to weep;
From the beardless lip, and the earnest eyes
We turn away to weep!

To weep for the life-long sorrow
Of his father's house and heart,
For the love that lives and lingers
When its ties are rent apart,
For the wounded soul whose sickness
Is beyond the healer's art!

To weep for the land we cherish,
So sadly and sorely cost,
For her noble sons that suffered,
For the best and bravest lost,
Defending her holy birth-right,
In the war's red holocaust.

5. One Union soldier/poet penned a few verses glorifying the raid on Huntersville.
Wheeling Daily Intelligencer, 1 February 1862

The Second Virginia, long rol[l] beat,
The Twenty-fifth Ohio come to meet,
Joined together as one did go,
Four miles, knapsack did throw.

We camped and did fires build,
Pickets were chosen to rebels heed,
And to cooking and we all did eat,
And we all laid down to sleep.

Next morning, all in good heart,
Major said we should not part,
We did go on till Mingo Flats,
Looking out for rebels' gray caps.

We traveled till we came to Big Springs,
The rebels been there was signs;
Still went on till the Blockade,
And looking at it around we parade.

The blockade of timber ten miles long,
The troops were singing a Union song,
And Indiana cavalry joined to fight,
Pickets we can run for spite.

Bloody Second and Twenty-fifth thought it fun,
And we made the minnie balls hum;
The rebels left like a blue streak,
Killed and wounded ten on their retreat.

Twenty-fifth to right to cut retreat,
Bloody Second to left in town beat;

Major Owens hoisted the colors,
The stars and stripes did wave of ours.

The boys of the Second cheered and said,
Taking of Huntersville it paid;
Some swords, pistols and guns,
And we all thought it fun.

And some got coats, some caps,
Others got boots, shoes and knapsacks,
Each man was laden down,
That seven hundred taking the town.

We could not get all away,
Set fire to the town it could not estray;
Eighty thousand worth was our gain,
So we left nothing there remain.

The soldiers all in the street cheering,
The boys in good hear nothing fearing,
And the rebels left and run,
Heard nothing of the rebels till Stanton.

Majors Owens andWebster are in command,
Two brave officers I know will stand;
The Bloody Second will stand till last,
Twenty-fifth has fought time past.

6. "Captain McNally's Boys"

Written by R. Robinson
Ironton Register, 27 February 1862

Do you remember Captain McNally's boys
And when they left Ironton?
Or do you ever get to hear
What they are doing—have done?

'Twas on, the twenty-third of June,
We left our native home,
For Camp Carlisle, Virginia,
Our Company to be sworn,

CHORUS – For there are as good men in it,
I really do believe,
To cut their way, Secesh to slay,
As ever were in the field.

'Twas on the twenty-seventh of June,
Which we remember well,
When Captain Craig came into camp,
All good, sound men to tell,
'Twas on the twentieth of July,
The first March we did make,
From Wheeling down to Benwood,
A railroad ride to take.

We rode from there to Webster,
And then we did encamp,
Leaving the railroad at that point,
Through the country to tramp.
Our march we did continue,
The battle-grounds to see
O'er Laurel Hill to Rich Mountain,
Then camped at Beverly.

To Elkwater Camp we then did go,
When we were called upon
To cross the mountains at daylight,
To the place called Beckey's run.
Fifteen days we did stay there,
Then back to camp again
Leaving a company of Zouaves,
That weak post to maintain.

The cesh were on the mountain side,
Full fifteen thousand strong,
And when they heard us coming,
They did not stay there long,
For they left there on double quick,
As you can plainly see,
For each man got a blanket,
And some got two or three.

The next attack that we did make
Was on Allegheny camp,
Where we came up in front and rear,
The enemy to flank.
Right before their breastworks
We then did form our line,
Over one hundred we did kill
In about eight hours' time.

The thirty-first of December,
Our spirits were very low,
Until we got marching orders,
Some fifty miles to go.
The first day of January,
The Big Springs we did make;
We did encamp there for the night,
Some refreshments to take.

We started the next morning,
Some thirty miles to make
Over hills and lofty mountains,
A Secesh town to take.
When we came to their outer post,
A desperate charge they made,
But we soon gave them to under-stand
That we had learned the trade.

We charged on them at double-quick,
And made them leave their ground,
And then we had a merry time,
In burning down their town.
Now Huntersville's in ashes,
It was conquered by the brave,
And on the top of the court house,
The stars and stripes do wave.

We are on Cheat Mountain summit,
To stand the wintry blast,
And company H, of bloody second,
Is guarding Cheat Mountain Pass.
Our friends will please remember us
We're fighting for them here
And send us all the news from home,
Our soldier life to cheer.

7. A soldier wrote this poem to describe picket duty in western Virginia, "performed in great, gloomy forests, with which the mountainous regions hereabouts are mainly covered. The picket post is usually on some obscure bridle-path away up in the mountain's side, or in the narrow ravine at its base, which divides it from its neighbor-hills, all equally elevated, precipitous and gloomy—and often-times miles distant from camp."
Written by E. H., Camp Elkwater, Virginia, October 14, 1861
Cincinnati Daily Gazette, 22 October 1861

A lonely spot! Dark forests dense,
For weary miles outstretch around,
And far the lonely path from hence
That echoes back the wagon's sound.

How monarch-like, leaf-crowned their forms,
Uplift those noble pine and ask
They know a hundred winter's storms,

But not the axman's ringing stroke.

A dreary night, nor moon nor star,
Scarce yield one ray to cheer the gloom;
Away from camp and comrade far,
The picket where may be his tomb,

The boughs o'erhead low-bending grow,
The moss beneath is old and green;
Amid the bushes crouching low
He peers, death-still from forth between.

His rifle rests upon his knee,
And on the stock two firm hands press;
Ah! Well he knows how cheerily
It heeds his finger's quick caress.

Three weary hours—or more—are gone;
The midnight must be drawing nigh;
The brooklet at his feet runs on,
He hears it's murmuring melody.

A soothing sound! He thinks of home,
Of loved ones left at duty's call;
And flocking round him there they come,
The same old faces, forms and all.

The grey-haired sire leans on his staff,
The matron—lives with God in heaven;
He hears his brother's ringing laugh,
His sister's loving counsel given.

But there is yet another still,
A girlish form of simple grace;
How beats his heart, his pulses thrill,
Still gazing on that trusting face!
Not long; a near, quick, startling crash,

And home, and friends, and all are lost,
As, where he looked for foeman's flash,
The prowling beast steals past his post.

The night wears on; a full hour more
Creeps drearily and slow away;
The moments pass the midnight hour,
And glide into another day.

The winds arise; he hears o'er head
Their wrestlings in the upper deep;
He knows to-night the Storm King dread
No common revelry will keep.

Long-echoed through these forest aisles,
The snuffling wolf his warning brays;
The answering cry, from distant hills,
The stealthy panther's haunt betrays.

The fitting night birds shrilly scream,
Defiant of the gath'ring blast;
With hollow roar and fitful gleam,
The storm around him bursts at last.

A fearful storm! The night is black,
The torrent pours, the tree-tops reel,
And, as it were dark doom's-day's wreck,
Red lightnings flash, and thunders peal.

Against his sturdy tree close prest,
The picket's dripping form is leant,
And though no shelter, it is rest;
Thank Heaven! the tempest's wrath is spent.

The quiv'ring leaves their showers distil,
The swollen stream sweeps madly on,

The north wind low is numbing chill
To him that weary waits the dawn.

It comes at last – Oh, beam of hope!
Thank God that doth the day restore;
The sun mounts up the eastern slope,
And comrades, camp, is gained once more.

INDEX

Abbott, Albert I. 48, 195
Addison 171
Albright, Christopher 180, 209
Alderman, Mr. 166, 207, 210
Alexander, D.W. 101, 144, 145, 146, 156, 157, 166
Allegheny Mountain, Battle of vi, ix, xi, xiii, 107-139, 141, 150, 185
Allegheny Summit xi, 11, 26, 46, 47, 61, 62, 66, 86, 87, 90, 91, 92, 93, 94, 96, 98, 100, 102, 105, 109, 110, 135, 136, 137, 148, 149, 161, 164, 165, 173, 176, 177, 178, 203
Ammen, Jacob 49
Anderson, Joseph 124
Anderson, Pierce B. 36, 39, 40, 124
Appomattox Courthouse 187
Arbogast, Henry 160, 166, 210
Arbogast, James C. 86, 173, 176
Arbogast's 57, 63, 67, 73
Arkansas troops:
3rd Arkansas Infantry 11, 16, 28, 36, 40, 56, 62, 64, 91
Army of the Kanawha 10, 83
Army of Northeastern Virginia 5
Army of Northwest Virginia 17, 52
Army of Occupation, Western Virginia 11, 54
Army of the Potomac 5, 105
Atkins, James 15, 29, 56
Augusta County, Virginia 162, 174
Augusta Rifles 117

Back Fork of Elk River 76
Badger, Joseph 186-188
Baldwin, John 26, 27, 46, 47, 61, 62, 105, 126
Baltimore & Ohio Railroad 1, 11, 16, 137, 154
Barbour, Lucian 49
Barbour County 34, 141
Barbour Greys 16, 181
Barksdale, C. G. 151, 154, 162, 163, 164
Barton, Seth 27, 51, 84
Bath Cavalry 145, 146
Bath County, Virginia 181, 182
Batts, William 119, 121, 129
Beem, David 48
Belington 141, 143
Bell, Horace 66-68, 70-74
Benjamin, Judah P. 61, 80, 82, 83, 89, 91, 98-102, 104, 133, 136, 139, 156
Bennett, William 77
Benwood 1, 250
Beverly 5, 10, 27, 55, 57, 78, 107, 109, 110, 154, 157, 163, 164, 167, 172, 174, 175, 179, 183, 250
Bidwell, Solomon J. 54
Bierce, Ambrose ix, xi, 48, 127, 195
Big Sewell Mountain 10, 19, 52, 190
Big Spring 15, 21, 23, 56, 144, 148, 248
Black Fork of Cheat River 141, 163, 172
Black Ridge 187
Blake, William H. 86
Blandy, B. W. 94, 167
Block Run 126
Booker, George E. 167
Boykin, Francis M. Jr. 4, 117
Bracken, James R. 110
Bracken's Cavalry 21, 110, 116, 123, 144, 145
Brady's Gate 75
Brockenbrough, J. W. 136, 137
Brooks, A. L. 130
Brown, Daniel W. 38
Brown, John 36
Bryant, Isaac 33
Buchannan, Mr. 78
Buckhannon 78
Buffalo Gap, Virginia 105
Buffalo Mountain 187
Buffalo Run 114
Busby, Barney 115
Buzzard, Eli 160, 210

Cammack, John Henry 42, 129
Camp Allegheny ix, x, xi, xii, 91, 92, 93, 97, 103, 104, 107, 109, 113, 114, 115, 116, 123, 124, 125, 128, 131, 132, 137, 139, 140, 141, 148, 149, 154, 155, 161, 168, 169, 173, 176, 177, 178, 180, 181, 183, 184, 187, 206
Camp Baldwin 187

Camp Bartow 8, 11, 13, 15, 16, 17, 19, 21, 25, 26, 27, 28, 29, 30, 34, 36, 37, 38, 39, 41, 42, 46, 47, 50, 51, 52, 54, 55, 56, 57, 59, 60, 61, 62, 63, 65, 66, 67, 68, 70, 71, 73, 77, 79, 83, 84, 85, 86, 87, 89, 90, 92, 93, 103, 110, 111, 113, 148, 183, 187, 195
Camp Elkwater 54, 55, 56, 74, 75, 107, 143, 144, 165, 252
Camp Foster 74, 78, 94
Camp Luther 172
Carnifex Ferry, Battle of 10, 19, 190
Cather, Fabricius A. 141
Cavins, Elijah H. C. 49
Centreville 78
Chapman, Abel 86-87
Charlotte Cavalry 28, 58, 85
Cheat Mountain xi, 6, 8, 10, 11, 12, 14, 15, 16, 19, 24, 48, 55, 57, 60, 63, 65, 82, 91, 93, 94, 100, 102, 108, 110, 143, 159, 164, 175, 180, 187
Cheat Mountain District 11, 50, 55, 107, 167, 178
Cheat River Bridge 98
Cheat Summit vi, xi, 8, 10, 11, 14, 15, 16, 19, 21, 26, 28, 29, 31, 39, 57, 60, 62, 63, 65, 66, 67, 70, 73, 83, 86, 87, 92, 94, 95, 103, 104, 106, 107, 110, 151, 154, 168, 172, 175, 176, 178, 181, 183, 196
Chewning, Jack 166, 210
Chilcott, John 125
Churchville Cavalry 47, 65
Cincinnati Daily Commercial 12
Cincinnati Daily Times 49
Circleville 173, 174, 183
Clinton, Wharton R. 121
Clover Lick 26, 176, 179
Colfax, Schuyler 105-06, 165
Collins, Mr. 57
Collins, Martin V. B. 120
Conner, Z. T. 135, 156, 168
Cool, Watt 77
Cooper, Samuel 66, 83, 89, 101, 102, 137, 150
Cooper, William P. 176
Coplin, Asa 143
Corley, James 58
Corrick's Ford 5, 68, 111, 140
Covington, Kentucky 91
Cox, Jacob D. 2, 19, 178
Crab Bottom, Virginia 17, 91, 148, 173, 176, 177, 184
Crooks, Robert Nelson 25
Crouch's 156, 180
Cumberland, Maryland 80
Currence, Jacob 105
Cutright, Matt 172

Dabney, C. E. 118
Daum, Philip 30
Daum's Battery 30, 31, 40, 41
Davis, General 159
Davis, Jefferson 8, 66, 76, 124, 181, 191
Dawson, Samuel 15, 60
Dedrick, Henry 162
Dedrick, Mary 162
Delay, Charles xi, xii, 96, 129
Delzell, Hugh 145
Department of Ohio 1
Deshler, James H. 127, 135
Dixie Boys 151, 154, 163, 164, 172, 173, 174
Doherty, Alexander 78
Donelson, Daniel S. 83
Dopp, Frederick 154, 206
Downs, George 181, 182
Dumont, Ebenezer 68, 69, 240
Dunn, McKee 106
Dry Fork of Cheat 58, 59, 140, 141, 143, 151, 154, 162, 163, 164, 167, 172, 173

Earle 176
Ebbon, Joseph H. 163, 164
Elk Mountain 144, 179
Elkwater 8, 10, 11, 21, 24, 55, 56, 75, 107, 110, 180
Elkwater Valley 75
Elza, Samson 141

Floyd, John B. 10, 15, 83
Fluornoy, David 163, 167, 173, 174
Ford, Thomas 30, 109
Forks of Waters, Virginia 16, 24, 26, 87, 148
Fort Pickens 76-78
Fort Pickens, Florida 74
Fort Washington 195, 239
Foster, Robert 23
Franklin 26, 27, 173, 174, 177, 178, 183
Franklin, Indiana 34
Fremont, John C. 182
Fulkerson, Samuel V. 60, 68, 73, 83, 84

Gandy Creek 151
Gardner, Jesse 175
Garnett, Robert S. 5, 8, 111, 189
Gate City Guards 38

Gauley Bridge 10
Gay's 166
Georgia troops:
1st Georgia Infantry 36, 38, 43, 86, 87, 91, 93, 197
12th Georgia Infantry 11, 18, 29, 36, 43, 46, 56, 60, 68, 79, 84, 87, 90, 92, 97, 113, 115, 117, 119, 129, 135, 140, 148, 155, 156, 161, 168, 206
Gibson, Samuel 143, 159, 166
Gillilan, M. B. 166
Glenn, Jefferson 103
Gordon, Jonathan W. 111-12, 130, 204
Gordon, Joseph R. T. 111-12, 125, 130, 204, 245-47
Gordon, Moses 115
Grafton 1, 4, 14, 80, 98, 109, 154
Grafton National Cemetery 195
Graves, Frank 40
Green, Henry K. 156, 206
Green Bank 8, 15, 16, 25, 26, 47, 51, 62, 67, 70, 83, 85, 86, 103, 111, 114, 167, 171, 176, 178
Green's 57
Greenbrier Bridge 60, 79
Greenbrier Bridge (Marlinton) 144, 145, 159
Greenbrier County 15, 83, 159
Greenbrier River 8, 12, 16, 27, 51, 63, 68, 91, 92, 110, 111, 113, 124, 140, 145, 175, 176
Greenbrier River, Battle of xiii, 30-53, 60, 119, 135, 155, 195, 242
Greenbrier Valley 33, 63, 66, 82, 85, 87
Greenland 92
Greenland Gap 174
Guhra, Francis 24
Gum, Joe 67
Gum, William 28
Gum Road 31, 33, 51, 70, 71

Hadley, John 14
Hall, James 25, 37, 48, 161
Hamilton, Robert 123, 127
Hamilton, William 33, 109, 115, 116, 123, 193, 201
Hampshire County 80, 166
Hansbrough, George 86, 91, 115, 116, 117
Hansbrough's Battalion 36, 38, 39, 91, 103, 115, 116, 117, 119, 120, 121, 127, 132
Harding, Stephen S. 130
Hardwicke, William W. 149
Hardy County 58, 59, 66, 80, 98, 103, 166
Harman, Michael G. 15, 65, 66, 148, 150
Harper, Bill 151
Harper, Jacob 151, 162, 164
Harpers Ferry 1
Harrington, George W. 76, 77, 121
Hart, Samuel 96
Hartman, Mr. 67
Harvey, George 123
Heavener, Uriah 67, 85
Heavener's 27, 57, 86, 89, 103
Heffren, Horace 23
Herbert, Lieutenant 63
Herrold, Jess 48
Hershee, Oliver P. 146
Heth, Henry 181, 182
Highland County, Virginia 24, 176, 179, 181, 182, 183, 186
Hightown, Virginia 128, 148, 149, 168
Hildreth, Dr. 127
Hill, John Lyon 28, 54, 57, 60, 65
Holmes, Albert 34
Holt, William 16
Howe's Battery 31
Howell, Joseph 24
Hull, Felix 16
Hull, George W. 85
Humphries, W. D. 49
Huntersville 6, 8, 26, 27, 83, 87, 100, 101, 103, 105, 143-150, 155, 156, 157, 159, 160, 166, 167, 175, 176, 177, 179, 180, 181, 184, 206, 207
Huntersville Raid vi, xiii, 143-150, 156-157, 159, 205, 249
Huntingden, Benjamin F. 184
Hurt, John L. 86
Huttonsville 6, 8, 11, 27, 55, 74, 75, 78, 107, 110, 133, 143, 144, 155, 156, 164, 166, 175, 178, 179, 180, 183
Huttonsville-Huntersville Turnpike 6, 8, 21, 36, 75, 144, 159, 175, 179, 180
Hyman's Battery 178

Indiana troops:
7th Indiana Infantry 11, 14, 29, 34, 43, 46, 49, 55, 68, 86, 87, 105-106, 107, 109, 110, 111
9th Indiana Infantry ix, 11, 19, 25, 28, 31, 33, 37, 41, 47, 48, 50, 55, 68, 70, 71, 72, 85, 114, 125, 127, 130, 155, 184, 239
13th Indiana Infantry 11, 15, 21, 23, 29, 41, 55, 74-79, 109, 110, 114, 115, 116, 120, 121, 122, 132, 201
14th Indiana Infantry 11, 19, 21, 24, 31, 32, 37, 39, 48, 49, 50, 55, 182, 243
15th Indiana Infantry 11, 19, 24, 29, 31, 55, 75, 78, 199
17th Indiana Infantry 11, 29, 31, 49, 54, 55, 199
Ingersoll, Frank 15, 41, 47, 132

Jackson, George 27, 105, 162, 163-164, 167, 173, 174
Jackson, Henry R. 8, 11, 12, 13, 15, 16, 26, 27, 36, 37, 39, 43, 46, 47, 49, 52, 54, 57, 58, 61, 62, 66, 80
Jackson, Thomas J. 80, 81, 83-83, 89-90, 98, 100, 102, 103, 104, 137, 139, 187
Jackson, William L. 40, 42, 157, 158, 159-160, 166-167, 168, 169, 174, 178, 181, 197, 206
James River and Kanawha Turnpike 1
Jeffers, Dennis B. 179-180
Johnson, Edward 36-37, 39, 41, 43, 54, 83-84, 86, 92, 93, 100, 102, 105, 110, 114, 115, 117, 118, 119, 120, 123, 127, 131, 133, 134, 135, 136, 137, 138, 139, 148, 149, 150, 156, 157, 159, 160, 163, 164, 166, 167, 168, 169, 173, 174, 175, 176, 177, 180, 181, 184, 202, 207
Johnson, Henry A. 121
Jones, A. C. 166, 175, 179, 180, 207, 209
Jones, Charles 70, 147
Jones, James A. 116
Juarez, Benito 66

Kanawha Valley 10, 57, 190
Kelley, Benjamin F. 104-105, 137
Kersh, Adam 161-162
Kimball, Nathan 21, 31, 32, 34, 42, 239
Kincaid, Colonel 183
King, James M. 121, 132
Kirkpatrick, Thomas M. 115
Kittle 176
Knapp's Creek 146, 160
Knight, Lieutenant 145, 146
Kuhl, Christian xii

Lantz, Joseph H. 154, 162, 163, 164, 172, 173, 174
Latham, George 141, 142, 143, 151, 154, 172, 173, 183, 184
Laurel Fork 59
Laurel Hill 5, 57, 68, 250
Lee, Robert E. 4, 6, 8, 9, 10, 15, 52, 98, 124, 176, 187
Lee, William Fitzhugh Henry 26
Lee Battery 124
Letcher, John 179, 182
Lewis, Andrew 124
Lewis County 74, 78
Lewisburg 178, 181
Lilley, Robert D. 117, 118, 129, 169-170
Lincoln, Abraham 6, 79, 96, 107, 165
Little Kanawha River 90
Little Levels 159, 175
Loomis, Cyrus 21, 40
Loomis's Battery 21, 31, 37, 39, 240
Loring, William W. 6, 8, 10, 15, 26, 27, 80, 82, 83, 87, 88, 89, 90, 91, 98, 100, 101, 102, 104, 105, 136, 137, 139, 150, 160, 180
Lucy, Jackson 46, 150
Lynchburg Artillery 36, 39, 92, 124, 149

Mackey, Sy 125
Madden, Thomas ix, 125
Manassas, Battle of 5, 8, 156, 175
Manassas, Virginia 5, 89, 90, 101
Marlin's Bottom 101, 144
Marrow, Ike 56
Marshall, Jacob W. 23, 145, 175
Marshall, Robert 144
Marshall's Store 23
Mason, C. R. 27

McCann, Nathan 39, 242-243
McClellan, George B. xiii, 1, 3, 5, 6, 8, 90, 105, 165
McCoy, Charles D. 128-129, 135-136
McCoy, F. G. 58, 59, 80
McDonald, Captain 164
McDonald, Isaiah B. 11, 116, 201
McDowell, Virginia 105, 168
McDowell, Battle of 155, 187
McDowell, Irvin 6
McGee, Bill 133
McIlvain, William 129
McLaughlin, William 146
Meadow Mountain 19
Middle Fork Bridge 78
Middle Mountain 6, 8
Millboro, Virginia 6, 8, 101, 103, 178
Miller, John 62, 126
Mills, Mr. 176
Milroy, John 70-71
Milroy, Robert H. 33, 42, 49, 51, 55, 65, 66, 68, 73, 106, 107, 108, 109, 111, 112, 113, 114, 116, 120, 121, 123, 124, 130, 131, 132, 133, 143, 150, 154-155, 164-165, 172, 173, 178-179, 180, 182-183, 184, 187, 197, 201, 239
Milroy, Samuel 107
Mingo Flats 56, 175, 176, 248
Mollohan, William 121
Monongahela National Forest xi
Monongahela River 90
Monterey, Virginia 6, 8, 16, 26, 65, 66, 80, 82, 83, 84, 85, 87, 89, 101, 103, 105, 136, 146, 148, 149, 150, 156, 167, 174, 176, 177, 178, 183, 184
Moody, Gideon C. 114, 116, 124, 125, 127, 132, 201
Moore, George E. 4
Moore, Henry 76
Moore, Jackson 175
Moore's 28, 57
Moorefield 58, 59, 82, 137, 163, 164, 177
Morgan, C. S. 39-40, 44-45
Morton, Oliver P. 11, 65, 74-75
Moss, John 184
Mountain Tigers 141
Mouth of Seneca 172, 174, 183
Muckalee Guards 119
Munford, Jack 37
Murphy, Daniel 122
Myers, A. C. 27

Nicely, George W. 28-29
Nicholas, A. F. 31
Norfolk, Virginia 1
Norris, Abraham 85
North Fork of Cheat River 172, 173, 174
North Mountain 174
Nottingham homestead 114
Nutter, John 118

Ohio troops:
- 3rd Ohio Infantry 199
- 6th Ohio Infantry 21, 23, 24, 54, 55, 56, 199
- 24th Ohio Infantry 31, 33, 34, 39, 49, 71, 72, 86, 199, 242
- 25th Ohio Infantry 33, 34, 39, 49, 67, 70, 71, 72, 104, 110, 115, 116, 120, 122, 132, 133, 143, 144, 146, 147, 150, 165, 183
- 32nd Ohio Infantry xii, 11, 19, 30, 31, 33, 46, 55, 70, 71, 85, 109, 110, 114, 115, 116, 120, 122, 123, 127, 131, 150, 154, 164, 167, 172, 175, 184, 201

Ohio River 4, 5, 11, 17

Parkersburg 1, 11, 14, 80
Pendleton County 58, 98, 103, 140, 141, 162, 173, 177, 181, 183
Pennington's 163, 164
Petersburg 16, 26, 27, 58, 87, 175
Philippi 5, 27, 34, 129
Phillips, Sylvester 171
Pierpont, Francis H. 165-166
Pinick, Jacob 120, 131
Pittsylvania Cavalry 65, 86, 92, 118
Pocahontas County vi, xiii, 8, 11, 12, 17, 19, 159, 160, 166, 178, 179, 181, 182
Pocahontas Rescuers 105, 145
Poe, David 40, 103, 116, 119, 121, 127
Poe, Solomon 38
Point Mountain 76
Porter, Jonathan G. 112
Porterfield, George A. 4, 5
Potts, J. N. 101, 144, 145, 146
Price, J. Urner 39, 243-245

Price, William 128
Prickett, Thomas 41, 93, 125
Pryor, Samuel 201
Pryor, Shep 18, 59

Quimby, George 171

Ramsey, James N. 36, 38, 86
Randolph County vi, 7, 11, 12, 109, 151, 164, 167, 175, 181, 187
Randolph Riflemen 145, 146
Redusky, Joe 42
Reger's Battalion 36, 115, 122, 128, 129
Reid, S.V. 28
Reminiscences of a Ranger 74
Reynolds, Joseph J. 11, 12, 19, 21, 22, 24, 29, 39, 42, 43, 46, 47, 48, 50, 51, 52, 55, 61, 65, 66, 68, 70, 74, 79, 91, 107, 109, 116
Rhoades, Charles 112
Rice, William 36, 40
Rice's Battery 38, 40, 57, 160
Rich Mountain 5, 250
Rich Mountain, Battle of 5, 68, 124, 189
Richards, Archibald 145, 146
Richardson, William P. 49
Richmond, Virginia 4, 6, 49, 58, 102, 155, 181, 206
Richmond Dispatch 79
Richmond Enquirer 60
Riddle, John 39, 242-243
Rigby, Silas F. 110
Riley, Mr. 89
Ringgold Cavalry 30
Rippetoe, William F. 203
Robinson, W. J. 117
Robinson's Cavalry 75
Rockbridge Cavalry 58
Rockingham County, Virginia 174, 182
Rocky Mountains 94
Romney 80, 82, 84, 87, 89, 90, 98, 100, 137, 138, 139
Rosecrans, William S. 6, 10, 19, 20, 51, 52, 55, 57, 61, 150, 164, 178
Ross, John D. 46, 110, 137, 140
Rowlesburg 154
Rust, Albert 10, 28, 36, 40, 56, 60, 63, 64

Salisbury Run 114
Schenk, Robert 55
Scott, Winfield 4, 5
Seneca 27, 58, 66, 140, 163, 164, 167, 173, 174
Shenandoah Mountain 177
Shenandoah Valley 6, 187
Shumaker, Lindsey M. 16, 36, 37, 39, 40, 46
Sickman, Alfred 125, 184, 185
Sill, Joseph 39
Simmons, David 163
Skeen, William 166, 181, 182, 210
Slaten's Ridge 114
Slaven's Cabin 15, 26, 28, 54, 57, 60, 63, 67, 75, 83, 86, 110, 111
Slayton, John 17, 48, 67, 114, 140, 166, 168, 176, 183, 184, 191, 203
Slayton, William 48
"Slayton's Retreat" 203
Slemmer, Adam G. 74
Smith, Charles E. xii, 175
Smith, Dan 42
Smith, Ed 42
Smith, John 42
Smith, Lewis E. 33
Snatchburg 78
Snider, Joseph C. 16, 92, 97, 118, 120, 176
Snyder, John 58, 59, 151
South Branch of the Potomac 59
South Branch Valley 58, 61, 80, 90, 178
Spaulding, Albert 94
Spirit of Democracy 96
Spriggs, John L. 181, 182
Spurgeon, George B. 180, 209
Starks 171
Staunton, Virginia xiii, 5, 6, 11, 14, 25, 26, 27, 61, 66, 82, 83, 87, 89, 90, 91, 98, 100, 101, 102, 104, 105, 109, 131, 136, 139, 148, 149, 154, 155, 165, 177, 178, 196
Staunton-Parkersburg Turnpike vi, viii, ix, xiii, 5, 6, 8, 10, 11, 14, 33, 49, 51, 61, 79, 80, 85, 87, 93, 102-103, 139, 171, 178, 182, 187, 189
Staunton Spectator 160
Stenbeck, John 175
Stephens, Alexander 135
Stephenson, Adam 136
Stevens, Sergeant 163
Stevenson, C. L. 26, 27, 102, 136
Stonewall Brigade 137
Stringtown 172
Sturges, Shelton 86
Sturm, Henry 34

Sullivan, Jeremiah 21, 23, 24, 56, 75, 76, 77, 241
Summermire, D. 25
Summit, Benjamin 65, 196-197
Sutton, George B. 203
Swiger, Dr. 75, 77

Taliaferro, William B. 36, 41, 91
Taylor, John 150
Tebbs, William H. 62
Templeton, B. F. 103
Thompson, Lewis S. 119-120
Thompson, Robert G. 85
Thompson, William P. 120
Tidler, Luther 40
Tolley, P. A. 101
Tract Bridge 58
Traveler's Repose 8, 39
Tucker County 140, 150, 163, 167

Upshur County 174
United States Infantry, 19th 130

Valley District 80, 82
Valley Mountain 8
Vanmeter, Joseph 80, 82
Varner's 97
Virginia, Restored Government of 166
Virginia Central Railroad 6, 8, 11
Virginia and Tennessee Railroad 178
Virginia troops (Confederate):
2nd Rockbridge Artillery 92, 126, 127
9th Virginia Infantry Battalion 36, 92
19th Virginia Cavalry 23
23rd Virginia Infantry 36, 41
25th Virginia Infantry Battalion xi, 11, 92, 96, 98, 117, 119, 129, 169
31st Virginia Infantry xii, 16, 25, 34, 36, 37, 42, 48, 68, 86, 92, 97, 98, 101, 105, 113, 115, 116, 117, 118, 119, 120, 128, 129, 145, 148, 157, 158, 173, 176, 178, 181, 206
37th Virginia Infantry 16, 60, 68, 73, 83, 84
44th Virginia Infantry 91, 128, 129, 160, 173, 203
52nd Virginia Infantry 11, 16, 24, 26, 27, 46, 47, 61, 62, 63, 86, 92, 105, 111, 126, 140, 155, 161, 162
Virginia troops (Union):
2nd Virginia Infantry 11, 55, 56, 75, 110, 114, 125, 132, 141, 144, 146, 150, 151, 165, 171, 172, 182, 183, 184, 185
3rd Virginia Infantry 143, 171, 179
10th Virginia Infantry 179
Virginia State Rangers 181

Walker, William 66
Warm Springs, Virginia 6, 27, 87, 100, 105
Washington, George 168
Washington, D. C. 1, 4, 5, 6, 96, 105
Washington University 11
Weaver, A. J. 151, 171
Webster 27, 55, 109, 154, 178, 250
Webster, George 143, 144, 145, 146, 147, 148, 183, 249
Webster County 74, 78, 171-172, 173
Weller, Enoch 72
Welton, Alf 50
Westbrook, Ulysses 115
Wheeling 1, 5, 14, 250
Wheeling Daily Intelligencer, 151
White, Soldier 141
Wilder, John T. 31, 49
Wilfong's 100, 168, 169, 176
Williams, Jeremiah 70, 146
Willis, Edward 164, 177
Wilson, Clayton 61
Wilson, Daniel 141
Winchester, Virginia 5, 10, 80, 82, 102, 104, 124, 137, 150, 177
Windy Run 180
Wolfenbarger, Robert 203
Wood, John T. 117-118, 129
Wooddell, William H. 112
Wooding, George Washington 37, 39

Yeager, Andrew 8
Yeager, Evelyn 94, 185
Yeager, Jacob 199
Yeager, John 93, 126, 199
Yeager, Mack 94
Yeager home 115, 126, 129, 185

Zanesville, Ohio 94, 127
Zanesville City Times 132

ABOUT THE AUTHOR

Joe Geiger, Jr. has published numerous articles and a book, *Civil War in Cabell County, West Virginia 1861-1865*. He is currently the director of West Virginia Archives and History, where he has worked for the past fifteen years. He also serves as an adjunct professor at Marshall University.